SEX DETERMINATION AND SEXUAL DIMORPHISM IN MAMMALS

ALFRED GLUCKSMANN
Cambridge

WYKEHAM PUBLICATIONS (LONDON) LTD
(A MEMBER OF THE TAYLOR & FRANCIS GROUP)
CRANE, RUSSAK & COMPANY, INC., NEW YORK
1978

Sole Distributors for the U.S.A. and Canada
CRANE, RUSSAK AND COMPANY, INC., NEW YORK

First published 1978 by Wykeham Publications (London) Ltd.

© 1978 Alfred Glucksmann
All rights reserved. No part of this publication may be reproduced, stored in a retrieval system, or transmitted, in any form or by any means, electronic, mechanical, photocopying, recording or otherwise, without the prior permission of the copyright owner.

Library of Congress Cataloging in Publication Data

Glucksmann, Alfred.
    Sex determination and sexual dimorphism in mammals.
    (The Wykeham science series)
    1. Mammals — Physiology.   2. Mammals — Anatomy.
    3. Sex — Cause and determination.   4. Sexual dimorphism
    (Animals)   I. Title.
QL739.2.G57   599'.03'6   78-63273
ISBN 0-8448-1370-2

Photoset by Red Lion Setters, Holborn London
Printed in Great Britain by Taylor & Francis (Printers) Ltd.,
Rankine Road, Basingstoke, Hants RG24 0PR

Contents

Preface	vii
Part I. Sex Determination	1
1. Sexual and Asexual Reproduction	3
2. Differentiation of Chromosomes	8
3. Mitosis and Meiosis	19
4. Gonadal Development and Sex Determination	30
5. Neuroendocrine Factors in Sex Determination	39
6. Sexual Maturation and Decline	48
7. Sex Ratio and Life Span	58
8. Sex Differences in the Rate of Growth and Maturation	72
Part II. Sexual Dimorphism in Adult Mammals	87
9. Differences in the Proportion of the Body and its Organs	89
10. Metabolism, Thermoregulation and Biorhythms	102
11. The Blood-vascular System and the Lymphomyeloid Complex	114
12. The Endocrine System	121
13. The Central Nervous System and the Sense Organs	130
14. The Skin and its Appendages	136
15. The Respiratory and Digestive Organs	141
16. The Urogenital Tract	146
Part III. Common Features in the Sex Determination and Sexual Dimorphism of Mammals	148
Abbreviations and Glossary	163
Reading list and reference books	171
Index	173

Preface

Differences in the biology of male and female mammals are not restricted to the reproductive organs and their functions. They exist on the cellular level, in the proportions of the body and its composition, in the relative size of the organs, the rate of growth, in metabolism and in the organization of the brain.

In female cells only one of the two X-chromosomes functions and this is selected at random. Female tissues are a mosaic of cells of which one half has an active paternal and the other an active maternal X-chromosome, while in males the single X-chromosome is derived from the mother and active in all cells. Faulty genes on the X-chromosome become manifest in males, while in females the defect occurs in only 50% of cells and is compensated for by the normal 50%. The distinctive chromosomal composition of males and females accounts for the prevalence in males of such congenital disorders as some forms of haemophilia.

Sex-linked structural and functional modifications are seen in many organs; for instance, the salivary glands, the thymus, the kidney, the olfactory epithelium in some species. The liver, spleen, heart, bones and teeth, muscles, fat depots, appendages of the skin are larger in proportion to the size of the body in males in some species, and in females in others. Sexual variations in the rate of growth, the rate of metabolism, temperature regulation, enzymatic activity, hormone levels, the composition of the blood, the competence of the immune system are reflected in the size of males and females, in their life expectancy, their sensitivity to environmental agents, their abilities to adapt and relate to the maintenance of health.

The fluctuations during the day in illumination and temperature influence the pattern of activity of all mammals and find their expression in the circadian rhythms. The sexual cycles of females modify the periodicity of these phenomena which are regulated by the neuroendocrine centres in response to external and internal stimuli. The differentiation of these centres in the brain varies with sex and can be altered by suitable treatment during a critical period of the development. The initiation and maintenance of cycles can be suppressed in genetic females and induced in genetic males. The circadian and the superimposed sexual cycles affect the metabolic processes and their rates, the sensitivity to stimulation and the reaction to drugs and poisons.

There are many gaps in our knowledge of the sex differences in human and animal biology. In laboratory animals, females may be preferred for investigations because they are easier to handle; or alternatively, males to avoid the fluctuations due to the oestrous cycles. Results are obtained frequently for only one sex and assumed to apply equally to the other. This practice is usually followed in the study of farm animals, where the more valuable female is investigated, while the male is castrated or killed early. In human physiology little is known, for instance, about the sex differences in the functions of the digestive tract and the role played by paracrines, prostaglandins and locally acting agents. The sexual dimorphism varies with the species: males are larger in some (men, bulls, stags, elephants, rats) and females in others (hamsters, rabbits, spotted hyena). Findings on one species can be extrapolated to another only with great reservations. Since some data cannot be obtained directly in investigations on men, some physiological features have to be interpreted in the light of animal experiments and observations on isolated organs and cells. Great caution is needed in the analysis and applicability of such data, and in generalizations based on them.

In all mammals, the differentiation and maintenance of the sexual dimorphism is determined by the varying interaction of chromosomal, hormonal and neuroendocrine factors. These affect also the psychosexual development of individuals; this aspect is dealt with in *Man and Woman, Boy and Girl*, by Money, J. & Ehrhardt, A.E. (1972) Baltimore & London, John Hopkins University Press, and in *Gender Differences: Their Ontogeny and Significance*, by Ounsted, C. & Taylor, D.C. (1972) Edinburgh, Churchill-Livingstone, and will not be considered here.

The sexual dimorphism in mammals is of theoretical and practical importance for the understanding of the biology under normal and pathological conditions and for the prevalence in males and females respectively of congenital abnormalities, of metabolic, infectious and neoplastic diseases. This monograph is the first attempt to present a systematic account of this rather neglected subject, though it cannot claim to be comprehensive or definitive. It aims to present facts rather than hypotheses and to stimulate interest in the problems and encourage research to fill in the many gaps.

This book is an expansion of an article in *Biological Reviews* (1974,

49, 423-475) which contains many relevant references.

My special thanks are due to Mrs Margaret Smith of Norwich for her enthusiastic and stimulating cooperation.

Many of my colleagues have provided information, discussion, criticism and suggestions and I am grateful to all for their help and in particular to Drs A.R. Dain, M.B. ter Haar, D.B. Stephens and D.E. Walters.

The award of an Emeritus Fellowship by the Leverhulme Trust Fund facilitated the preparation of this book.

PART I
SEX DETERMINATION

1. Sexual and asexual reproduction

The lifespan of all living organisms is limited, and they reproduce to survive as such in the case of unicellular forms, or in others to persist to some extent in their offspring. Protists split into two identical copies after a period of growth, and repeat this process as long as suitable conditions of nutrition, temperature, oxygen tension and the absence of toxic substances permit. The genetic constitution, structure and physiology of bacteria remain constant for countless generations and capable of variation only within very narrow limits. Probably since their creation every species has maintained its complement of genes arranged along a single chromosome and some additional plasmids composed of DNA (deoxyribonucleic acid). It is constituted by identical and largely invariant entities.

A change in the genome of a single member is a rare event, which is unpredictable and occurs at random as the mutation of a gene or as the unidirectional transfer of part of a chromosome during the conjugation of two identical members of the species, or by the incorporation of a plasmid from another related or unrelated protist. A mutation is the permanent alteration of the chemical structure of one or more genes. During a conjugation some genetic information is passed on from one protist only, and the recipient eliminates an equal amount of DNA. Competence to deal with bactericidal agents can be acquired by the incorporation of a plasmid or short length of a chromosome from a resistant related strain or unrelated protists. Thus the information necessary for the production of enzymes capable of inactivating a toxic substance is handed on, and allows a mutant or resistant form to overcome unfavourable environmental conditions. A series of strains and substrains differing in the capacity of making some enzymes thus develops, but the invariance of the species persists relatively unaltered, and the potential for evolutionary progress is restricted.

The unicellular protists have a single dose of genetic material, i.e., they are haploid; they are prokaryocytes, i.e., they have no nuclei in which the genome is separated from the cytoplasm by a nuclear membrane; they are identical copies of one another, i.e., they are asexual; they have a short span of active life and split at intervals of minutes or hours, though they are capable of survival for long periods in the inactive form of spores.

The multicellular vertebrates, including mammals, reproduce sexually, i.e., two individuals pass on their genetic information to each offspring, which is thus not an identical copy of either parent. The descendent is diploid, having a double dose of genetic material arranged in pairs of chromosomes which are enclosed in a nucleus; it is eukaryocytic, the nuclear membrane separating the genetic material from the cytoplasm. To avoid an unmanageable increase in the number of chromosomes by the contributions from both parents, the complement in the germ cells has to be halved, that is made haploid, from the diploid or polyploid state of the somatic cells of each individual. The reduction divisions, meioses, in the maturation of spermatozoa and ova serve this purpose as well as facilitating a further mixture of genetic information by the interchange of segments between homologous chromosomes before reducing them to the haploid conditions. Thus before their fusion during fertilization the genetic information in each germ cell is derived from the grandparents and further back in the line of ancestors. This two-fold process of interchange of genes, at first during meiosis and afterwards during fertilization, increases the variation between individuals of one species and thus the chances of producing some organisms best suited to adapt to changes in the environment, and so to evolutionary progress. Germ cell production is different and separate from that of somatic cells, and is restricted to the gonads, the testis and the ovary.

The combination of gametes with different genetic information is possible in theory without sex difference, the differentiation of male and of female germ cells in male or female gonads. Animals, unlike plants, are not sessile and do not depend on intermediaries such as bees for the transport of pollen. The gametes have to traverse great distances relative to their size and one of the pair has to be motile. This requires the shedding of all unnecessary cytoplasmic material, the concentration of the genetic material into a small package and the development of locomotory organelles. The early embryo requires nutrients for growth, and the other germ cell enlarges by the accumulation of food stores. The maturation of spermatozoa and of ova meets these requirements. Both could be and are produced by the same individual of some genera, which are hermaphrodites possessing ovaries as well as testes. Self-fertilization prevents the interchange of genetic material between two individuals, and safeguards against such a possibility exist, even in plants.

The evolution of the gonads in vertebrates and their restriction to individuals of one or the other sex has been slow, and a variety of developmental stages is seen in some genera and species. Some teleosts, bony fish, are true hermaphrodites with ovaries and testes present in every individual, though incapable of self-fertilization. Some adults can change from the female state in which the ovary functions predominantly, to the male state when ovarian activity is suppressed and the testis is active; the active sex is changed from female to male, when for instance the number of fertile adult males is reduced. True sex reversal occurs also in amphibia; all frog larvae raised at relatively high temperatures develop into females, and even adult male newts treated with the female sex hormone oestrogen turn into fertile females producing offspring, and may revert later to the fertile male state. In fish as in amphibia both gonads persist in an actually or potentially functional state. In all vertebrates, the gonads originate in a ridge of the mesonephros, the ontogenetic and phylogenetic precursor of the kidney, and the peripheral region develops into an ovary and the centre into a testis. Both parts develop in hermaphrodites, while in birds and mammals either the periphery, the cortex, or the centre, the medulla, differentiates to a functional state. This is one reason why sex reversal in these genera is at best incomplete. The other reason is the differentiation of some chromosomes with the specific function of determining the sex of the offspring; these sex chromosomes are often recognizably different, heteromorphic, from the others, which are called autosomes.

The chromosome complement of male and female gametes differs in having either two equal sex chromosomes and being homogametic, or having two different sex chromosomes and being heterogametic. In birds, the female sex is heterogametic and has a W- and a Z-chromosome, while the male has two W-chromosomes. In mammals, the male is heterogametic and has an X- and a Y-chromosome, while the female has two X-chromosomes. The heterogametic sex, like the homogametic, contributes only one of the sex chromosomes to the conceptus, and the sex is determined by the female bird, which imparts either a W- or Z-chromosome, while the male always imparts a W-chromosome. If the female contribution to fertilization is a Z, the offspring will be ZW and a female, but if the female contributes a W the offspring is male and WW. In mammals the male contribution determines the sex of the conceptus: a Y-chromosome of the sperm added to the X-chromosome of the ovum results in a male of XY constitution, an X-chromosome in an XX constitution, which is a female.

The development of sex chromosomes represents a further step in the evolution from asexual to sexual reproduction. Early stages of heteromorphic sex chromosomes can be seen in some fish as, for instance, the rainbow trout. The successful sex reversal of adult fish suggests either that the coexisting ovaries and testes produce sexually dimorphic gametes, or that in the sex determination the sex chromosomes are still subordinate to some of the autosomes. In birds, sex reversal is at best incomplete; in female chicks only the left gonad develops into a functional ovary; if it is removed from a newly hatched chick the right gonad, which normally atrophies, becomes a testis and produces spermatozoa. These are, however, heterogametic and carry either a Z- or a W-chromosome, unlike those of the normal cock which are always W in type. Thus the gonadal structure has changed from an ovary to a testis, but not the chromosomal constitution of the spermatozoon. The evolution of sexual reproduction involves at least two levels, the gonadal and the chromosomal, to which will have to be added as a third determinant the neuroendocrine level. In the heterogametic as well as the homogametic sex one of the two sex chromosomes finds its way into one of the mature gametes during the reduction division, though abnormalities are known to occur because of incomplete separations resulting in XXY, XXX or XO constitutions.

The ova are always considerably larger than the sperms. The egg of an ostrich weighs millions of times as much as a spermatozoon, which is of microscopic dimensions. Amongst mammals the eggs of monotremes have a diameter of 2·5 to 4 mm, while the head of the spermatozoon which carries the essential genetic information measures about $4 \times 2\cdot5 \times 1$ μm. Even in eutherian mammals, where the embryo is provided for by the placenta, the sperm is minute compared with the egg. The accumulation of food in the ova has to cover only the short period before the implantation of the fertilized ovum and the development of the placenta which is linked to the maternal circulation. The enlargement of the egg is due to the storage of nutrients and to the concentration of the cytoplasmic contents of four precursors into only one mature ovum, and the elimination of the nuclear material of the other three as polar bodies in the meiotic divisions (see below and fig. 2).

Apart from producing gametes the gonads manufacture the sex hormones which govern the proliferation and maturation of germ cells, the formation and maintenance of the structures of the genital tract, i.e. the tubes, uterus, vagina, vulva and clitoris in the female and the epididymis, prostate, penis and accessory glands in the male.

The testis produces predominantly androgens, the ovary oestrogens and progesterones, all of which are steroid compounds and related to one another. Their action is not restricted to the genital tract proper, but influences the skin with its appendages (hairs and mammary glands), the configuration of the pelvis, the growth of the body, the development of muscles, the deposition and localization of fat, and a great variety of metabolic processes. The most important function is the interaction of the gonadal hormones with the brain and in particular with the neuro-endocrine centres of the brain stem and its neighbourhood, the pituitary and pineal glands. They are in a feedback relationship essential for the regulation of the reproductive as well as the somatic functions of the body, of the sexual and circadian cycles, the responses to environmental factors and the initiation as well as the maintenance of sexual dimorphism in the structure and function of mammals. The role of the chromosomal or genetic, of the gonadal or endocrine, and of the central neuroendocrine factors in the determination of sex and sexual dimorphism is the subject of the following chapters.

2. Differentiation of chromosomes

The techniques for the analysis of the structure and functions of chromosomes have been developed and greatly improved in recent years. Morphological observations can be checked experimentally by genetic manipulations such as in-breeding and cross-breeding in animals, and by the study of the pedigree of people with an inherited abnormality. Thus a convincing picture of the role of chromosomes and of segments within them in the development and the physiology of organisms is being built up. All interpretations in this rapidly expanding field of science are liable to change in the light of findings following the introduction of new techniques or the improvement of old ones. The number of chromosomes in man was established as 46 only some 20 years ago, while previously many cytogeneticists had counted 48. A simple improvement in the preparation of the material for microscopic analysis has revolutionized the field; treatment of the material in a hypotonic solution before spreading it on a glass slide, instead of merely squashing it between glass slides, leads to a better separation of the individual chromosomes and more detailed study. This is helped by the use of ordinary or fluorescent stains, after enzymatic digestion, to reveal banded structures along the length of individual chromosomes, which facilitate their identification and also that of any defects in them. Tritiated thymidine is taken up by the chromosome during the synthesis of DNA and autoradiographic examination allows us to distinguish between segments of the chromosomes engaged in DNA synthesis and those that are not, at the same time. Immunofluorescence microscopy of markers on chromosomes, and the specific binding of some such stains to the Y-chromosome, have added further to the arsenal of cytogenetic techniques. The electron microscopy of stained sections of chromosomes and the surface appearance in the scanning electron microscope contribute further to the analysis of the details in the structure of chromosomes.

These cytological techniques are usually applied to cells which have been isolated and kept in artificial culture media to which growth-stimulating compounds are added to promote mitosis. Subsequently addition of agents such as colchicine or its derivatives arrest the cell division in metaphase. Thus the number of cells in phases suitable for detailed study of the chromosomes is greatly increased, and the search for the right stage of the dividing cell eased. Other new methods use biochemical assays of enzyme activity, or refined fractionation either following the fusion of cells of different origin into heterokaryons or following the hybridization of the cellular DNA with known DNA sequences from viruses; all these have proved to be powerful tools in molecular genetics.

The practical application of such studies to the screening of compounds as harmless therapeutic agents, as cancer-producing carcinogens, as teratogens causing malformations, as enzyme poisons or generally as toxic substances with isolated cells *in vitro* is rapid, relatively cheap and fairly reliable. It can be used only as a preliminary sifting of compounds into harmless and injurious, since in the body their action is modified by the circulation of the blood, by excretion and detoxification processes, and by compensatory actions of other tissues and organs which are absent *in vitro*. Thus experiments on animals are needed for a more reliable assay.

A very valuable role of these techniques is their help in estimating the risk of a congenital abnormality in the offspring by means of a study of the chromosomes of the prospective parents in genetic counselling before starting a pregnancy, or during it to detect an abnormality from the foetal cells of the amnion. Such material may be studied morphologically or by the assessment of the activity of specific enzymes in the amniotic fluid. In patients with inherited abnormalities of enzyme activity, cells from various tissues can be analysed to discover whether they are present in the same individual or in male or female relatives. Examples will be given later for such cases in man. Immunological studies can also be made on various types of cells for the presence of antibodies, for their capacity to respond to antigenic stimulation, and also for histocompatibility between a donor and a recipient of a tissue or organ graft.

In animals, the correlation between visible changes in chromosomes or activity of their genes and the appearance of distinctive features in the phenotype, such as colour of the hair coat or obesity, are tested in breeding experiments. For this purpose small laboratory animals such as mice are inbred for generations by brother-sister mating, or by back-crossing of the offspring with one of the parents. Such 'pure lines' usually have distinctive and easily recognizable features, and by hybridization with a member of another pure line the inheritance of the character can be investigated and the presence of modifying factors in the mother or in the genome of either parent studied. Small animals are used because they have a short generation time and the effects soon become demonstrable. They are also less costly to keep than larger animals. Visible characteristics such as the colour of the hair or eyes and the length of the tail have been used for a long time. Nowadays refined biochemical and immunological techniques for the analysis of gene action and inheritance are available and widely used. Maps of the localization of genes on a chromosome are based on the linkage of inherited characters; the more frequently two features appear together in the same individual, the closer to one another are the responsible genes. Translocations of some segments from one chromosome to another can be detected by this method and related to morphological changes in the chromosomes concerned. We shall be concerned mainly with the sex chromosomes in the analysis of sexual dimorphism in mammals and examples will be cited below.

Inbreeding in man occurs in some isolated communities, and some characteristics may appear as inbred. The isolation may be due to geographical conditions, to social factors, or to religious beliefs. An extreme form of brother-sister mating was practised in the families of the Egyptian Pharaohs. A study of such groups provides information limited by the numbers present and by the time scale of the generations. Pedigree studies of families in which a member manifests an inherited abnormality have been most useful in the study of man. This applies to defects such as haemophilia, colour blindness, various forms of mental deficiency and also to some congenital abnormalities which may have a genetic component as shown by their frequent manifestation in members of a family. We shall concentrate on differences with sex in the appearance of these defects, which are usually due to genes on the X-chromosome. The majority of gene defects are recessive and can be compensated for by a homologous gene, an allele, on the other chromosome of the same pair. Thus in females the other X-chromosome provides a safety valve for the manifestation of the faulty gene, while in the male, with only the one X-chromosome, no such remedy is available. The phenomenon is illustrated in the pedigree of some families. The best known example is that of haemophilia, which as a mutation was handed on by Queen Victoria through her daughters to her grandsons of the ruling families of Russia and Spain. None of her daughters or granddaughters manifested this defect, which is due to a recessive gene (h) carried on the X-chromosome present in both male and female offspring. It is manifest only in the males, but obscured by the normal allele on the other X-chromosome in females. The mutant allele is recessive, and thus does not appear in the daughters of affected males. Not all sons of women carrying the gene are affected, because the ova may contain either the normal or the faulty chromosome. The mother is heterozygous in respect of the gene. If the gene is faulty on the other chromosome as well, i.e., the mother is a homozygote in this respect, all the offspring manifest the deficiency since there is no compensatory effect. If such an allele is dominant, males and females are affected equally, provided they receive the faulty X-chromosome from the mother, even if she is a heterozygote. If she is a homozygote, the defect is manifest in all male and all female offspring.

The example also shows that the mere presence of a gene does not necessarily lead to its appearance in the phenotype of the carrier. This requires the presence or absence of a faulty allele on the other chromosome as well; it depends on the balance between them and also on other factors in the genome which give rise to the phenomenon of 'incomplete penetration of the gene'. It may also become operative only under certain metabolic or hormonal conditions. Thus baldness at the crown and temple is inherited by both men and women, but becomes manifest early on in men and only after the menopause in women. It needs for its effect levels of androgens which are present in males from puberty on, but are neutralized in women during their reproductive period by the female sex hormones. When the latter drop in relation to the androgens normally present in the female, baldness ensues. This phenomenon is described as sex-limited inheritance, i.e., it is carried on autosomes, but requires certain sexual characteristics for its appearance. In this example, it is the androgen level. Inheritance carried by factors on the sex chromosomes is termed sex-linked.

Some important cytological discoveries preceded the correct assessment of the chromosome numbers in men and the introduction of refined methods. One such observation is that of the 'Barr body', which appears in the interphase nuclei of the females close to the nuclear membrane as a distinct body staining deeply with nuclear stains such as the Feulgen stain, which is characteristic for the presence of DNA. This body is present in females of some species, including women, and not in normal males. A similar body is recognizable in the nuclei of leukocytes, usually referred to as the 'drum-stick' because of its resemblance to it. It also is restricted to females. Both structures have been identified subsequently as one of the X-chromosomes rendered inactive by heteropycnosis, i.e. condensation of the chromosome. These structures are not obvious in all female cells, but are always absent from normal males. They can be used for the diagnosis of the genetic sex of an individual, and also for the multiplication of the X-chromosome; in trisomy three

X-chromosomes are present, but only one of them remains active and there are thus two Barr bodies visible. Alternatively, females with XO constitution, monosomy of the X-chromosome, like males do not have a Barr body. On the other hand males of the XXY constitution have a Barr body.

Sex chromosomes and autosomes

The diploid number of chromosomes of mammals varies from 12 in some marsupials to 78 in the dog. It is 46 in man, 42 in the rat and 40 in the mouse. Obviously the number of chromosomes is not related to the size of the animals, nor is it to the total amount of DNA, since the individual pairs of chromosomes vary widely in length. The X-chromosome is not the largest, nor is the Y-chromosome the smallest of them, though in some species they differ by a factor of 3 or 4. From data on *Escherichia coli* and the fruit fly *Drosophila*, it has been estimated that in man about 100 000 genes are distributed over the chromosomes. About 30% of the DNA exists in multiple copies, and much of it must serve functions other than the coding for amino-acid sequences of the proteins. The X-chromosome in man accounts for about 6% of the length of the haploid chromosomes, and 93 loci are now known. If the same relation of number of loci to length of chromosomes applies, there should be about 1550 autosomal loci, of which at present 1050 have been discovered. These figures may have to be modified as cytogenetics advances. In 1962 only about 60 loci were listed for the X-chromosome, so their number had increased by 50% in 1975. The Y-chromosome, considered as a mere dummy, apart from its role in initiating the development of the male gonad, is now known to carry some histocompatibility factors in mice and probably in man.

Chromosomes are not merely strings of genes in a haphazard order, but have definite and individual structures, as evidenced by the centromere as connection with the spindle apparatus and the centrosome. The centromere is not necessarily situated in the centre of the chromosome dividing it into equal arms. In many chromosomes, short and long arms give a distinctive configuration to them. The structural genes for the coding of amino-acids are arranged in sequence with regulator genes, which control the activity. While the codes for amino-acids appear to be universal, their sequences in building up the proteins vary with species and, within them, with individuals. This is made evident by the specific immune reactions, by which the species and individuals can be recognized. The difficulties of matching tissues for transplantation of such organs as kidneys, liver, heart and skin are notorious, and even when they are approximately compatible, the host has still to be treated with immunosuppressive drugs to prevent rejection of the graft, at least initially.

The simple model of gene activity and its regulation elaborated for the haploid prokarycyote *E. coli* is not sufficient to explain the changing pattern of gene regulation in diploid eukaryocytes. For these, a model involving sensor genes which receive a stimulus and integrator genes to transcribe them has been formulated. This system assumes that a repatterning of genes in new combinations is responsible for the evolutionary steps. Some of the DNA not involved in coding may be employed in this process. The concepts of the action and regulation of sequential groups of genes are still in flux, and are likely to be enlarged and modified by progress in molecular genetics. It is quite evident that in the differentiation of various tissues and organs with their specific proteins only some of the gene complexes are active at a time, while others may remain quiescent for prolonged periods. That they can be reactivated is demonstrated by some experiments on amphibia; isolated nuclei from differentiated cells, of for instance the gut, transferred to enucleated eggs are able to supply the genetic information for the development of embryos from the hybridized egg cell. Human tumours produce hormonal substances or develop hormone receptors which are only latent in the parent tissue from which they are derived.

Chromosomes and segments of the same chromosome differ in their activity and the timing of DNA synthesis. Inactive chromosomes, or parts of them, tend to be condensed and 'late-labelling'. Tritiated thymidine, a labelled precursor of DNA, is incorporated during the synthetic phase and detected by autoradiography. A solution of the marker is applied, the tissue is fixed after various intervals, and then processed for histology. The specimen is covered by a photographic solution and developed. Silver grains are localized over chromosome regions which synthesize DNA at the time of fixation of the cells. If present at short intervals between the application of the labelled precursor and fixation of the tissue, they are called 'early-labelling' regions, while those becoming positive after longer intervals are termed 'late-labelling'. Condensed regions of chromosomes are usually late-labelling, and thus indicate a functional difference in the timing of DNA synthesis.

Autosomes are present as duplicates of paternal and maternal origin. Homologous though not identical loci of the pair on the whole act synergistically, though dominant alleles prevail by definition. Both members of a pair of autosomes are necessary for the functioning and survival of the individual, and this is proved by some abnormalities, as will be seen below. For sex chromosomes the position is different; the Y-chromosome is absent from normal females and thus obviously not necessary for the survival of the individual. Males have only single X-chromosomes and are thus able to function without the second member of the pair. The difference in length of the male and female sex chromosomes in man makes it unlikely that they carry the same amount of genetic information, and suggests that they differ mainly in their sex-determining functions. The Y-chromosome is usually heteropycnotic and is in a condensed state which indicates an inactive phase. Of structural genes on the Y-chromosome of man, only a rather doubtful connection with the appearance of 'hairy ears' in some families has been reported, and there is the recently discovered link with some histocompatibility factors. There is the possibility that the genes for these factors are actually located on either the X-chromosome or one of the autosomes and triggered into activity by the presence of the Y-chromosome. On present evidence the main function of the male sex chromosome is the initiating action for the development of the testis.

For the much larger X-chromosome almost 100 loci are known at present. Since males manage with only one of them, the duplication in females seems to be wasteful, as all the essential physiological processes are fundamentally similar in both sexes. The monosomy of the X-chromosome in males suggests that its genetic dosage is equivalent to that of a pair of autosomes of comparable length, or, put differently, that the dosage of genes of a single X-chromosome is double that of a single autosome of the same length. This phenomenon is referred to as 'dosage compensation' of the X-chromosome for its monosomy. In turn this has consequences for the female pair of X-chromosomes; since the male receives one from his mother and his sister retains the same chromosome, the X-chromosomes must be equivalent, and one of them is superfluous and has to be inactivated. This is indeed the case; one of them becomes heteropycnotic and appears in the nuclei of female cells as a Barr body or as a drum-stick in leukocytes. The inactivation of either the maternal or paternal X-chromosome in most eutherian mammals occurs at random, and M.F. Lyon has proposed and substantiated this hypothesis. Her investigations are concerned with the hair colour in mice of inbred strains. By cross-breeding males with females of different hair colour with genes on the X-chromosome, she has shown that the male offspring inherits the gene on the maternal X-chromosome and has a hair coat of uniform colour resembling that of the mother. In the daughter the colour of the hair coat and even of individual hairs is patchy, with some regions resembling the colour of the mother and others that of the father. This mosaicism of the hair colour is interpreted as random inactivation of either the paternal or the maternal X-chromosome in different clones of cells of the hair, or of hairs in different parts of the skin. The hypothesis states in simplified terms that this randomization in the inactivation in the female of one of the pair of the X-chromosomes occurs in embryonic development and remains constant in the offspring of these cells. A great deal of additional evidence for this hypothesis and the phenomenon of 'lyonization' of female cells has been adduced.

A few examples concerned with X-linked recessive abnormalities in man may suffice to illustrate this phenomenon. In hypohidrotic ectodermal dysplasia, manifested only in men, the sweat glands of the finger tips and the palm are missing; in female relatives of such individuals the total number of glands is reduced, but varies between patches with normal numbers and others without any. This variation in women, like that of the hair colour of mice, is a sign of the random inactivation of the paternal or maternal X-chromosome in various regions of the skin; inactivation of the faulty one results in normal development of glands, while that of the normal one leads to their absence. Another sex-linked abnormality shows itself as an abnormal development of the central nervous system, and has been traced to a defect in the gene responsible for the enzyme hypoxanthine-guanine phosphoribosyl transferase (HGPRT). Skin fibroblasts tested in tissue culture for the activity of this enzyme show that it is absent from all cells of males, but present in about half the cells taken from their female siblings and absent from the other half. Similar tests on lymphocytes and erythrocytes show no activity in males, but activity in all cells of female relatives. This contrasts with the findings in their skin fibroblasts and suggests that in some tissues the cells can respond to the faulty chromosome by preferential inactivation, or that cells carrying the active faulty chromosome are eliminated. A third example is that of the X-linked agammaglobulinaemia of boys, which lowers their resistance to infections because their lymphocytes cannot differentiate to plasma cells and produce the globulins that are needed in the formation of antibodies. In female siblings two types of lymphocytes are found in culture; those able to produce globulins and form plasma cells, and those which lack this capacity. The distinction again depends on whether the paternal or maternal X-chromosome is heteropycnotic.

These examples provide supporting evidence for the hypothesis of random inactivation of one of the X-chromosomes in the female, and for the thesis that only one is required for the survival and formation of a normal organism. The monosomy of the sex chromosomes seen in the absence of the Y in females, the absence of a second X in males, and the inactivation of one of the pair of X-chromosomes in the female, contrasts with the deleterious and often fatal effects of monosomy of autosomes. The organism can manage with only one X-chromosome, but needs both members of the pair of autosomes. Similarly, duplication of an autosome leads to severe anomalies if not early death. This condition is called trisomy and in the instance of Down's disease (formerly called mongoloid idiocy) chromosome 21 is present in three instead of the normal two copies. The effect is visible in a number of distinctive features, and also in the increased liability to contract certain diseases. Duplication of either of the sex chromosomes does not interfere with the survival of individuals, whether of males with the constitution XYY or XXY, or of females with the constitution XXX or XXXX. The additional Y-chromosome may lead either to increased gonadal development resulting in excess production of androgens and its consequences, or for reasons as yet not understood, to infertility. Duplication of the X-chromosome is dealt with by inactivation of the superfluous numbers and the appearance of two or more Barr bodies. Thus the inactivation of sex chromosomes (and the Y is normally heteropycnotic and inactive) is a normal procedure, and is extended to coping with additions to their number. The organism does not have this capacity in the case of the autosomes and thus additions to their number result in severe abnormalities.

Some exceptions to the random inactivation of an X-chromosome have been reported. In kangaroos cross-breeding between strains with distinguishable X-chromosomes shows that the paternal X-chromosome is always inactivated. This observation suggests some differences between the paternal and maternal X-chromosome quite apart from their different genetic content. In all mammals there is a generation gap due to the process of meiosis, which will be discussed in the next section. Some modification in this process in some species leads to some unusual somatic constitutions. In the creeping vole (*Microtus oregoni*) males are as in most species XY, but females have always XO somatic cells. In the wood lemming (*Myopus schisticolor*), females indistinguishable in appearance and in fertility may have the somatic constitution of either XX or XY. It is assumed that an X-linked gene in XY females represses the male determining effect of the Y-chromosome. In most other species, the presence of a Y-chromosome leads to the development of male gonads irrespective of the number of X-chromosomes, all but one of which can be and are inactivated. It is still not decided whether the male-determining function of the Y-chromosome is due to a direct action, or whether its presence triggers off gene activity on the X-chromosome, or even of the autosomes, resulting in the initiation of testis development. In sheep, some of the autosomes of the male are distinctly larger than their homologues in the female, and may thus have additional information for the male development. Whether a direct action or a triggering effect is involved, the presence of a Y-chromosome is usually associated with the induction of a testis and of male characteristics—except in the wood lemming. Obviously gene action depends on that of other genes on the same chromosome and on interactions between genes and chromosomes of the genome; these are modified in evolution and thus lead to the evolvement of strains or species.

Apart from the exceptions of XY female wood lemmings and the inactivation of the paternal X-chromosome in female marsupials, the somatic cells of female mammals are a mosaic, with active paternal or maternal X-chromosomes, while male somatic cells are uniform in always having the active maternal X. Examples of the expression of the female mosaicism have been given for the hair colour in mice and some inherited abnormalities in man. The first example shows the positive manifestation of it; the second shows its importance by compensating for and thus obscuring the gene defect of half the cell population. How far the mosaicism of female cells is responsible for the differential reaction of males and females to a variety of external stimuli, such as carcinogens or agents causing infectious and other non-neoplastic diseases with a differential sex ratio, remains to some extent a subject for conjecture. It is important to note that male mammals are heterogametic and produce male or female-determining sperms, while the females are a mosaic as regards the activity of paternal and maternal X-chromosomes in somatic cells.

The inactivation of one of the X-chromosomes in females occurs in early embryonic development, and species differ in the exact stage at which heteropycnosis is effected. Once established in a stem cell, its descendants perpetuate the heteropycnosis of the same X-chromosome throughout development and postnatal life. There is no information on how this process is ensured. It is possible that this chromosome, once rendered inactive, remains subordinate to the dominant active partner, and that this character is not changed during the process of cell proliferation. This may be due to different patterns in the timing of DNA synthesis of which there is evidence in the timing in early and late labelling parts. So far no obvious differences between the members of a pair have been found. The percentage of female cells showing a Barr body has been reported to differ in various tissues, and in the same tissue at various phases of the oestrous or menstrual cycle. These observations cannot distinguish between inactivation of the paternal or maternal component. The evidence from enzyme studies on female cells of siblings with manifest X-linked defects suggests that in some tissues a selection against the faulty gene, or rather chromosome, is practised, as shown in the comparison of skin fibroblasts with lymphocytes and erythrocytes in the case of HGPRT.

3. Mitosis and meiosis

The object of mitosis of somatic cells is the production of two equal and identical copies, while in the meiosis of the gonocytes the number of chromosomes is halved by two sequential divisions, while the amount of cytoplasm is largely shed in the formation of sperm or unequally distributed in the formation of ova. Thus the mitotic process is similar in essence to the asexual reproduction of prokaryocytes, while the meiotic process reflects sexual reproduction in the exchange between parts of chromosomes in a long drawn process prior to the actual separation of the two daughter cells. In both types of division two distinctive phases are essential; the first is the synthesis of DNA (S phase), in which the amount of genetic information carried in the chromosomes is doubled, and the second, called karyokinesis, is that in which equal halves of the genetic material are sorted out and separated. These two phases were formerly considered to be simultaneous, the duplication of the genetic material occurring at the same time as the separation of the chromosomes. Autoradiographic studies with labelled precursors of DNA, and measurements of the DNA content by other methods, have established that the DNA synthesis precedes the separation of chromosomes. For the cytoplasmic material, division may precede synthesis, as indicated by the size of the cell volume at the end of division and its subsequent growth to the volume of the parent cell. The processes of synthesis and separation differ in timing, and also in the mechanism of organelles involved. The nuclear constituents are the genetically important structures and are concerned with synthesis, while in the separation or packing of the cellular materials, the spindle apparatus, with the centrosomes and their link to the chromosomes by the centromeres, is involved. Since the rather striking movements in the second phase have been well documented in descriptions in text books and in films, they will not be dealt with here. A description of some aspects of mitosis, which is the same in cells of the two sexes, is given here to explain the differences between the two sexes in the process of meiosis.

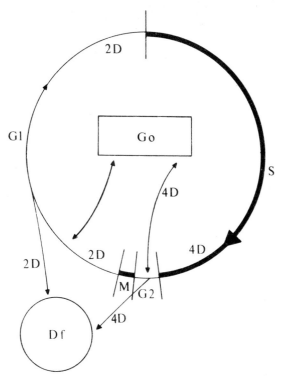

Fig.1. Phases of the cell cycle: the periods of actual mitosis (M) and of DNA synthesis (S) can be determined by direct microscopic observation or in autoradiographs. The two intervening periods (gaps) are called G1 in the presynthetic interval and G2 in the post-synthetic one. These periods vary considerably in duration and the intersections are only approximate indications of the time taken for each of the 4 periods. The proliferating cell proceeds in a clockwise direction from G1, which it enters with the DNA content of a diploid cell (2D), which is doubled in the S period to 4D and halved by M to 2D. Of the 2 cells formed in M, one will proceed via G1 to the cycle, while the other may leave it permanently for differentiation and functional activity (Df) or temporarily for Go. In this way the total number of cells of a tissue stays constant and the cell in Go is available for regeneration when necessary. If the cell number of a tissue is to be increased, both cells formed in M will enter G1. Cells may also leave the cycle as tetraploids in G2 for either Df or Go.

3.1 *Mitosis*

Autoradiographic studies have been predominant in showing that the period of DNA synthesis (S) is followed by a postsynthetic gap (G2), before mitosis (M) and separation of daughter cells, and this in turn by a presynthetic gap (G1) before the next cycle of synthesis and mitosis. The two gap periods are intervals between two distinct processes, but do not imply inactivity of other cell organelles. Growth of cytoplasmic volume occurs in G1, and changes in the physico-chemical state of the cytoplasm take place in G2. While the cycle thus described (fig. 1) is typical for the cells of a tissue as long as they remain in the proliferating compartment, some cells leave it and take up permanent differentiation (Df), which renders them incapable of replication, or reversible differentiation (Go), when they return after intervals of varying duration to the proliferative state. Even while proliferating, the cells may not complete the cycle. The S period may not be followed by nuclear division nor the latter by cell division. By this means tetraploid single cells or multinucleate cells are produced, and some of these are quite frequent in some organs, e.g. binucleate cells in the liver, multinucleate cells in the bone marrow and spleen, and mononucleate giant cells as ganglion cells. Cells differ in volume with the tissue, ranging from the small lymphocytes to the large ganglion cells. Some of these differences are due to special cytoplasmic activities, such as the accumulation of fat and sebum, the production and secretion of mucin and other secretory granula, the production of keratin and fibrils etc. These functions are carried out on the instructions issued by the genome in response to stimuli received via the cell surface and cytoplasm. Cells with large nuclei have as a rule a DNA content representing a multiple of 2n, and suggest that even when the S phase is not followed by that of M, the DNA content is doubled. When cells leave the proliferating compartment, they can do so in G2 with twice the DNA content of a diploid cell, or in G1 with the normal diploid content.

The interval between one mitosis and the next is referred to as the cell cycle time. The total duration of the cycle varies with tissue and stage of development between about 8 and 28 hours for developing tissues. The variations are largely in the duration of G1 and G2, while M is fairly constant between 30 and 60 minutes. The S period lasts between 5 and 12 hours. The generation time in the retina of a 2-day old rat has been estimated to last about 28 hours, divided into 13 hours for the presynthetic period (G1), 12·5 hours for the synthetic phase (S), 1·5 hours for G2 and 0·8 hours for the actual process of mitosis (M). This is only one example of many studies of the process reporting variations in the overall duration and particularly in that of G1 and G2. This is so particularly in cells capable of resuming division after prolonged periods with specialized functions. The cells of the rat liver may perform their metabolic functions for periods of months or even years in the Go period, but return to the cycle if partial hepatectomy requires increased regenerative activity.

The cells in the reproductive cycle adapt their activities to external as well as internal conditioning. A circadian cycle, i.e. one that lasts about 24 hours, with peaks of mitotic activity has been known for a considerable time, and is found to vary with tissue. This variation is ultimately linked with the photoperiod, the sequence of light and dark periods changing with the seasons, which regulates the activity of animals adapted mainly either to daylight or to nocturnal activities. These determine the time of feeding and the subsequent sequential stimulation and activity of the digestive apparatus, and the various regulatory mechanisms in associated organs such as production of insulin or of corticosteroids. It is of interest that such a circadian rhythm persists even in organs isolated from the body. It is still noticed in adrenals kept in a culture medium. Superimposed on it in females is the oestrous or the menstrual cycle, when in a complex feed-back relationship between the ovary and the pituitary-hypothalamic activity the cells of the female genital tract, and in particular of the uterus and vagina, start to produce masses of cells and subside afterwards into quiescent periods. Experimentally such activity can be induced by injecting immature female rodents or ovariectomized adults with suitable doses of sex hormones (oestrogens). The hormone elicits a wave of mitotic activity. For isolated blood cells in culture media stimulation of mitotic activity by the addition of substances such as phytohaemagglutinin is a widely practised method for cytological analysis.

The cycle is thus not simply a 24-hour clock which, once started, proceeds automatically, but depends on a variety of interactions between various organs, and varies in them and with environmental conditions such as feeding habits in relation to climatic conditions and the photoperiod. There are also feed-back regulations within a tissue, as for instance in the epidermis where the removal of the keratinized layer with adhesive tape is followed by a burst of proliferative activity in the deeper or basal layers. In this example the duration initially of G2, and subsequently of G1 is shortened until the equilibrium between the cell contents of the basal and superficial layers is restored. The G2 period of cells in the cycle is reduced, and the G1 of cells about to enter it is decreased, both effects contributing to the burst of mitosis. It should be remembered that following M (fig. 1) there are two daughter cells available for G1, and in functioning tissues one of them may leave the cycle to undertake specialized tasks, while the other remains in the proliferative compartment. If the total number of cells is to remain constant, there has to be a feed-back mechanism between cell proliferation and cell loss, either by cell death or by an intermediate step into the differentiating compartment.

3.2 Meiosis

In somatic cell division the sex chromosomes and autosomes behave in the same manner; they double their DNA content, split and separate. In the reduction divisions of gametes an exchange of segments between the members of a chromosome pair takes place, and while this is no problem in the female cells with an XX constitution, in male cells no pairing for the X or the Y chromosome is available, and they have to be separated from one another for the production of either X- or Y-bearing sperm. The exchange of segments between homologous chromosomes assists the aim of sexual reproduction, i.e. the creation of variability of individuals within a species, not only by randomizing the paternally and maternally derived chromosomes between the offspring, but by creating through interchange entirely different chromosomes from those inherited. The chromosomal components remain unchanged and thus the invariance of the species is maintained, but they are recombined in various forms on the same structural chromosomes. These can then be given in various combinations to the daughter cells. This is of particular importance in the female where only one mature ovum is formed by each oogonium instead of four spermatocytes by each spermatogonium. Cell division is the only way of redistributing the chromosomes, and since their number has to be reduced during this process the mixture of chromosomes is considerably restricted. Hence the recombination of components of chromosomes before the reduction in their number is of great importance for sexual reproduction and ensuring variability in the offspring, and thus enlarging the range available for selection.

One of the differences between mitosis and meiosis is the fact that in somatic cells the chromosomes replicate and form identical copies for the resulting offspring; in meiosis new chromosomes are formed by recombination of parts, and these are different from the parental chromosomes. Since the sex chromosomes of the male mammal are not present in homologous pairs, they remain unaltered by exchange and recombination in meiosis, unless they pair with one another in those segments not concerned with the determination of sex. This has been thought to be the case for a long time. Recent cytological observations on rodents and man suggest that at most an end-to-end contact takes place between the X- and the Y-chromosome in meiosis, and there is also no genetic evidence for an exchange between them. If the absence of structural genes on the Y-chromosome (with the exceptions mentioned above—sex determination, histocompatibility factors and those for hairy ears) is accepted as correct, exchange of parts between the sex chromosomes could serve no useful purpose, since no suitable alleles are available. Very recently reports—so far unconfirmed—suggest that about half the human Y-chromosome is made up of tandemly repeated DNA sequences which are considered to be genetically neutral and are not present in females. These sequences are believed to build up a largely inactive chromosome structurally to prevent pairing and chiasma formation for the exchange of segments. If an interchange between X- and Y-chromosomes occurred, some at least of the loci known on the X-chromosome should be detectable genetically as being linked to the Y-chromosome, and this has not been possible so far in rodents or man.

The failure of the sex chromosomes to pair in the male has the consequence that he transmits unchanged the X-chromosome he received from his mother. The maternal X-chromosome in females is a new combination due to the exchange between the mothers paternal and maternal X-chromosomes. Thus the paternal sex chromosome of a daughter is derived unchanged from her grandmother, while the maternal one is a new combination and made up by parts from her own mother and her paternal grandmother, i.e., it extends over two generations in a new combination. This suggests that there is a gap of one generation between the paternal and maternal X-chromosomes, and that the latter is more liable to carry changes than the unchanged paternal one. How the male effects the separation of the X- and Y-chromosomes in meiosis will be discussed below.

The duration of the meiotic process in the two sexes differs very considerably. Even in the male the premeiotic S phase is lengthened, and instead of 7 hours in the early mitotic divisions of spermatogonia it takes about 14 hours for the late cells in the spermatogonial sequence, and about the same time before the first meiotic division. In the late spermatogonia, and subsequently in the meiotic divisions, the separation of cells (cytokinesis), is incomplete and cytoplasmic bridges connect the daughter cells, ensuring some measure of synchrony right up to the differentiation of spermids (spermiogenesis). These connections may be responsible for the appearance of multinucleate precursors of sperm in the testis in individuals suffering from malnutrition or after exposure to drugs, toxins or irradiation.

Following the preceding S phase the first meiotic division is a lengthy and complicated process which involves the pairing of autosomes and the exchange of parts. It is usually subdivided into five stages, of which the longest is the pachytene, which lasts about 8 to 10 days in the male mouse and about 16 days in man. The total duration of both meiotic divisions in the mouse is about 13 to 16 days, of which most is spent in the first division, while the second division is fairly

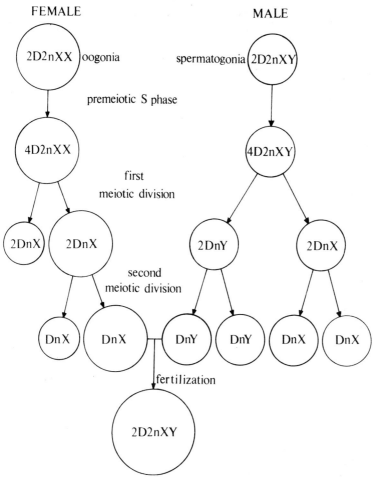

Fig.2. Meiotic divisions in male and female mammals. Oogonia and spermatogonia enter the reduction divisions as diploid cells as regards DNA content (2D) and chromosome numbers (2n), and of these a pair may be either XX or XY. These cells are thus 2D2nXX in the female and 2D2nXY in the male. During the premeiotic S phase the DNA content is doubled, and the cells are then 4D2nXX or 4D2nXY. In the male the sex chromosomes form the XY body indicated as XY. During the following first meiotic division the chromosome number and DNA content are halved, and the X-chromosome separated from the Y-chromosome and two cells formed as either 2DnX or 2DnY in the male, while in the female unequal cytoplasmic division produces a large cell and the first polar body, both 2DnX. The second meiotic division follows without an intervening S period, and halves the DNA content but keeps the chromosome number and distribution the same. In the male four cells are produced as either DnY or DnX, and in the female one large ovum and by unequal cytoplasmic division the second polar body, both DnX. On mating a haploid sperm with a haploid egg a diploid fertilized ovum of either 2D2nXX or 2D2nXY is obtained.

fast. Since mitosis of somatic cells takes about 45 minutes the duration of the first meiotic division exceeds it by about 400 times, which can be taken as an indication of the complexity of the process. In both cases the period of the preceding S phase has been omitted; it is known to vary by a factor of 2.

The two meiotic divisions are necessary (*a*) to reduce the number of recombined chromosomes to one half, (*b*) to redistribute them equally as regards the autosomes over two daughter cells and (*c*) to reduce the DNA content of the whole genome to one half. These purposes are achieved in two steps which can best be understood with the help of a diagram (fig. 2). This represents the sequence of events in the female on the left and in the male on the right, and neglects the complicated convolutions of the chromosomes for the exchange of segments by chiasma formation which are illustrated in all text books. In the premeiotic S phase, like that of a somatic mitosis, the total amount of DNA is doubled: if D2 describes the condition in a diploid cell, it will be D4 at the end of the S phase. Hence the condition of spermatogonia at this time can be described as being 4D 2n (the diploid chromosome number) and XY with both sex chromosomes: 4D2nXY. After rearrangment of the chromosomes, the first meiotic division produces two cells of which each contains 2D, but half the chromosome number n, and either an X- or Y-chromosome: 2DnX or 2DnY. This means that though the number of chromosomes has been halved, each still contains double the amount of DNA. During this process the sex chromosomes are separated from the autosomes and form an XY body, which has also doubled its DNA in the premeiotic prophase. The XY body remains separate from the autosomes while they pair and exchange parts, and rejoins them only for the process of separation in diakinesis. Thus the first step has achieved the halving of the chromosome numbers, the recombination of the autosomes, and the separation of the sex chromosomes into two cells, but still leaves all chromosomes with double the amount of DNA for a haploid cell, which has to be halved in the next step. Hence the second meiotic division follows without an intervening S phase, and equal splitting of the chromosomes results in the formation of two cells each of DnY or DnX constitution. There are now four haploid spermatocytes with the correct haploid amount of DNA and separation of the X- and Y-chromosomes.

The meiotic divisions in the female serve the same purpose as those in the male, but having two X-chromosomes for pairing they need not be separate from the autosomes during these phases of the first meiotic division (left of fig. 2), which results, however, by unequal division of cytoplasm, in a large cell with most of it and a small one (the first polar body) with very little; both have the same nuclear constitution of 2DnX. Only the large cell matures further, and undergoes the second meiotic division without a prior S phase but again with unequal cytoplasmic division leading to the extrusion of the second polar body. Thus instead of four spermatocytes with about equal cytoplasm but differing in either having an X or a Y-chromosome, only one large ovum is formed which contains the cytoplasmic volume of four cells. At fertilization the combination of a sperm with an ovum leads to the normal constitution of a diploid organism: 2D2nXY or 2D2nXX.

In most mammals, oogenesis is completed in the foetal period, i.e., no further oogonia are formed by mitotic division; this is the condition in man, mouse, rat, guinea-pig, sheep, cow, pig and monkey. In these species the number of oogonia reaches a peak before or at birth and subsequently declines. In the rabbit and cat some multiplication of oogonia takes place in the neonatal period. At this stage the premeiotic S phase has been completed in all these species, and the cells enter the prophase of the first meiotic division and remain in that condition until puberty—for 12 to 15 years in girls. From that time onwards hormonal stimulation leads at about monthly intervals to the formation of larger follicles around the ovum, to ovulation and shedding of the ovum with its surrounding cells into the genital tract, and to completion of the first meiotic division with formation of the first polar body. The ovum enters immediately into the second meiotic division, but does not proceed unless stimulated by the entry of a spermatozoon at fertilization, when it extrudes the second polar body. The meiotic process in women may thus last about 50 years, from birth to menopause, although it is arrested for at least the period between birth and puberty. Furthermore, completion of the meiotic divisions depends on the further stimulus of the entry of a spermatozoon.

In the male, multiplication of spermatogonia and their differentiation into fertile sperm starts at puberty and continues throughout life. The process of sperm formation in the testis has a wave-like pattern, and in man takes about 74 ± 5 days from the initial proliferative stage to the final differentiation of the spermatozoon. There is no diurnal rhythm in spermatogenesis, and no cycles like the oestrous or menstrual cycle of females. It is a continuous process, and through a number of mitotic as well as meiotic divisions results in the formation of 96 spermatozoa by each stem cell. In women 1 ovum is produced in each follicle at monthly intervals, and up to 18 in the mouse at 4 to 5 day intervals, unless pregnancy supervenes to stop ovulation. Thus there are enormous differences between the sexes in the production of germ cells as regards numbers, rhythm and time scale.

The following figures illustrate the differences in the production of germ cells in men and women: during a reproductive period of about 38 years the stock of 500 eggs initially available is reduced at each cycle, i.e. about 13 times per annum, and not replenished. An ejaculate contains on average 3×10^8 spermatozoa which are produced continuously. Assuming 2 ejaculations per week as an average over 50 years and full replacement of the sperm count, the total production of spermatozoa is 15×10^{11} as a rough approximation. The estimate of a daily production of 1×10^8 sperm in the testis produces a figure of the same order for a period of 50 years. At successful matings the chance for a sperm to fertilize an ovum is and remains throughout life 1 in 3×10^8. A sperm count of 2×10^7 is considered the minimum required for male potency, as most die in the female genital tract and release enzymes which facilitate the migration of a minority and the penetration of the mucus and of the corona of cells surrounding the ovum. The wastage in the production of germ cells will be discussed in the section on gonadal differentiation.

The long duration of the first meiotic division is due largely to the sorting out of chromosomes after pairing and exchanging segments, and is fraught with difficulties which may result in some abnormalities of pairing and exchanging parts, as well as of separation. Thus translocations of segments of chromosomes, or of some parts of them, can be recognized by cytogenetic studies, and failure of separation, called non-disjunction of chromosomes, gives rise to trisomy or monosomy. Trisomy of chromosome 21 in Down's syndrome can be due either to translocation of part of one of the chromosomes to the other of the pair, or to failure of separation. Non-disjunction of the sex chromosomes at this stage leads to constitutions resulting after fertilization in XXY or XYY configurations, or in XXX and XO. Nor are these failures restricted to members of a pair of chromosomes. In the mouse a reciprocal translocation between the X-chromosome and autosome 16 is characterized by male sterility, and in females by the lack of mosaicism of the X chromosome (Searle's translocation). Another example is that of Cattanach's translocation of part of autosome 7 and the X-chromosome.

So far only chromosomal events affecting sexual differences prior to fertilization have been discussed. The spermatozoon contributes mainly the genetic information contained in the DNA of its head, which is formed by the pycnosis of the nucleus. The ovoid spermatocyte is transformed into a tadpole-like or sickle-like sperm according to species. Some of the cytoplasmic constituents, together with parts of the cell membrane, form a cap or acrosome around the head, while others differentiate into a middle piece and into a tail to ensure the sequential release of enzymes and the motility of the sperm. Some cytoplasmic substances are discarded as Regaud's bodies, which contain no nuclear material in contrast to the polar bodies in oogenesis. Spermatogenesis is not complete with the reduction to the haploid state, and some of the transformations of the cell to a mature sperm take place in the epididymis and even after passing into the female genital tract, and are steps necessary for the successful penetration of the corona around the ovum and of the egg.

4. Gonadal development and sex determination

The two aspects of gonadal functions are the production of germ cells and the production of hormones in the interstitial tissues; they are reflected in their ontogenesis. The gonads, without the primordial germ cells, start as the genital ridges on the mesonephros, the Wolffian body, which is an intermediate stage in the development of the kidneys. The gonocytes are first seen in the yolk-sac endoderm near the caudal end of the body of the embryo. They appear on the 8th day of pregnancy in the mouse, and migrate via the mesenchyme of the mesentery into the genital ridge on the 10th day, completing the migration on the 12th day.

At these stages of development the gonads are described as indeterminate or indifferent, because there are no apparent differences between the future testis and ovary in gross morphological appearance. The cells forming the gonads are determined by their sex chromosomes as either XX, and so female, or XY, and so male. The male cells multiply faster than the female ones, and in the rat embryo of 14·5 days the volume of the male gonad is 40% greater than that of the female. At this time more germ cells are present in the male than in the female gonad. Two distinct regions can be distinguished in the gonad at this stage; the peripheral zone or cortex and the central region or medulla, which is closer to the Wolffian body. The primordial oogonia settle in the cortex and primordial spermatogonia in the medulla. Fig. 3 indicates roughly the localization of the germ cells in the gonads, and also that two duct systems are associated with them. The Wolffian duct is that of the mesonephros, and the Mullerian duct is formed by the folding of a groove on the genital ridge. In both sexes both ducts are present, though their subsequent development differs. The Mullerian duct has an open end cranially, and terminates caudally in the urogenital sinus, which encloses also the end of the Wolffian duct and that of the ureter coming from the definitive kidney, the metanephros. The Mullerian duct forms the Fallopian tube which is close to the ovary, and continuous with it the uterus and the upper part of the vagina. The Wolffian duct together with some adjacent mesonephric tubules forms the epididymis, the ductus deferens, the seminal vesicles and at the urogenital sinus the prostate and accessory glands. Most of the unwanted parts of the gonads atrophy and are retained merely as rudimentary structures: the cortex becomes the tunica albuginea, the tough capsule of the testis; the mesonephric tubules of the medulla the epoophoron and paroophoron, and remnants of the Wolffian duct form the female prostate, and of the Mullerian duct the utriculus masculinus. Older embryos can be sexed by the size of the gonads and the presence of central tubules of peripheral follicles.

Various explanations are offered for the faster growth of the male than of the female gonad. The Y-chromosome is smaller and has less DNA than the X-chromosome, and thus may be able to synthesize DNA more quickly and the cells may duplicate faster in the male; there is as yet no supporting evidence for this idea. Another explanation is based on the well-documented growth-promoting effect of testosterone in the adult, and assumes that this steroid exerts a similar action on the gonad. The secretion of androgens by the interstitial cells starts on the 14th to 15th day of pregnancy in the rat, and reaches a maximum value on the 18th day when it reaches a plateau. The secretion thus starts in the gonad of the rat before the immigration of gonocytes ceases.

The gonad develops independently of the presence of primordial germ cells, as is evident from the time-relations during normal ontogenesis, and is also supported by experimentation; if the gonocytes are removed by surgery or by localized irradiation before they start their migration, the gonads proceed with the normal differentiation of the interstitial tissues, though they remain sterile. In a strain of mice (W/W) an inherited abnormality in the migration of germ cells results in the formation of normally secreting gonads in the absence of germ cells in homozygous individuals. The female mule is a sterile cross between a he-ass and a mare, but has an ovary with persisting secretory functions.

An isolated guinea-pig testis kept in organ culture starts to secrete androgens on the 22nd or 23rd day of the gestation period. An isolated ovary begins secretory activity under the same conditions only on the 41st to 46th day of pregnancy, with the production of androgens presumably as an initial step to the biosynthesis of oestrogens. The Mullerian ducts of guinea-pigs do not seem to respond to either the female or the male sex steroids in culture, since they develop alike whether the ovary or the androgen-secreting testis or adrenal gland is present. This observation cannot be extrapolated to other species, to other stages of development or to different doses of

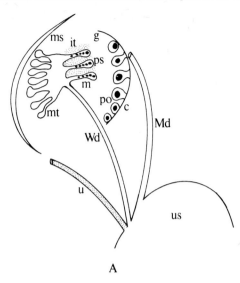

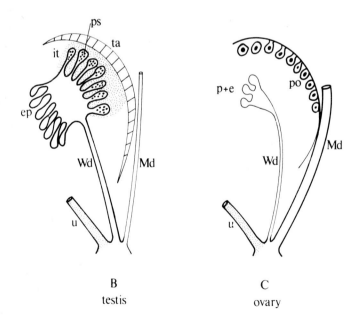

B
testis

C
ovary

androgens (see below). In many mammals the early secretory activity of the testis is responsible for the persistence in the males of the Wolffian ducts, the differentiation of the epididymis, the seminal vesicles, the prostate and the accessory glands as well as for the development of the external genitalia. Testosterone secretion by the foetal testis inhibits the development of the mammary glands, and is partly responsible for the regression of the Mullerian ducts. Other testicular secretions appear to be involved as well in the regression of the Mullerian ducts, since androgens alone fail to suppress the formation of uteri from the Mullerian ducts, while testicular implants do so.

The role of testicular secretions in the determination of sex is illustrated by the appearance of freemartins in ruminants such as cattle and goats, which have been known to farmers for a long time. A freemartin is the female of a bisexual pair of twins with normal external genitalia but with sterile ovaries. The cortex of the ovary forms a tunica albuginea like that of a testis, and the medulla forms seminiferous tubules; the differentiation of the ovary is inhibited and masculinized. The cells of the body remain of the XX constitution. The change in the development of the female gonad is due to the presence of vascular connections (anastomoses) between the placentas of the male and female twin. These allow the testicular secretions to come into contact with the developing ovary, to upset the normal differentiation of the cortex and to transform the cranial parts of the

Fig.3. The contributions of the mesonephros, the foetal gonads and the Wolffian and Mullerian ducts to the development of the foetal testis and ovary.

A *The mesonephros in association with an imaginary hermaphroditic gonad.* The mesonephros (ms) consists of the Wolffian duct (Wd) and the mesonephric tubules (mt). The gonad (g) is formed by the cortex (c), where the primordial oogonia (po) settle, and the medulla (m) in which the primordial spermatogonia (ps) assemble in cords surrounded by interstitial cells (it), and link with the mesonephric system. In mammals only male or female germ cells reach the gonad and either the cortical or the medullary regions develop. The Mullerian duct (Md) is formed at the periphery of the gonad, and joins the urogenital sinus (us) together with the Wolffian duct and the ureter (u) of the metanephros.

B *The foetal testis* develops in the medulla and connects with the Wolffian duct (Wd) and the mesonephric tubules, some of which persist as the epididymis (ep). Ample amounts of the interstitial tissue (it) are present and in a secretory state. The cortex atrophies to form the capsule of the testis, the tunica albuginea (ta), and the Mullerian duct (Md) regresses.

C *The foetal ovary* forms in the cortex, where the oogonia (po) multiply and develop into primary follicles. The interstitial tissue around the follicles is sparse and inactive. The Mullerian duct (Md) persists to differentiate into the Fallopian tubes, the uterus and part of the vagina. The Wolffian duct (WD) and the mesonephric tubules regress to the rudimentary structures of the paroophoron and epoophoron (p + e).

In B and C persisting structures are indicated by bold and atrophic by faint outlines.

Mullerian ducts. In rats, rabbits and marmosets such intersexes do not appear in spite of anastomoses between the placentas, because enzymes in the latter are able to metabolize the androgens; these enzymes are not present in the ruminants. Androgens alone cannot account fully for the formation of freemartins, since an injection of androgens into the pregnant cow allows the female foetus to retain her ovaries, but causes masculinization of the external genitalia. The testis may have another active compound, as demonstrated in the rat by the different effects produced by testicular grafts and by androgen injection on the regression of the Mullerian duct. The additional 'factor X' or 'medullarin' acts only over short distances, and its effect diminishes with distance from the testicular implant.

Anastomosis between the placentas of twins allows circulation both of primordial germ cells and of blood cells between the members of the pair. They are thus cellular chimaerae of XX and XY constitution. In the male calf, XX cells in meiosis have been found in the testis, but they are destroyed during development as are the XY cells in the female. Only female gonocytes reach meiotic prophase in the foetal period, male ones fail to do so and are dormant until puberty. Thus it is unusual to find XY cells in meiosis in the female twin. Chimaerism of germ cells does not account for the formation of freemartins, where the genetic (chromosomal) and hormonal (gonadal) sexes diverge, producing sterile intersexes with essentially female external genitalia, but also some male sexual characters. In some of these intersexes, ovarian as well as testicular rudiments persist. There is evidence that the interstitial tissue of both the male and the female gonad is capable of converting the common base (cholesterol) into either the male or female type of sex steroids. Thus secretion of testosterone by the gonads of XX intersex goats and cattle has been reported, and the secretory cells are XX Leydig cells instead of the normal XY cells of the testis.

Antiandrogens are compounds developed from progesterone, one of the female steroid hormones. In adult males they cause atrophy of accessory sex organs such as the seminal vesicles, the prostate and various sebaceous glands. In pregnant rats they affect the male foetuses by regression of the Wolffian ducts, the development of a female type of urethra and of external genital organs, the formation of mammary glands with nipples, the induction of female cycles of activity in the hypothalamus and pituitary, and a female type of sexual behaviour. In pregnant dogs they induce the regression of the Wolffian and Mullerian ducts in the foetus, and promote the formation of a vagina from the urogenital sinus. Early castration of male foetuses causes regression of the Wolffian ducts, but not of the Mullerian. In rabbits foetal castration leads to the disappearance of the Wolffian ducts, and to the development of the Mullerian ducts and of accessory organs of female type. Thus the effect of castration and of the administration of antiandrogens varies with species and with the stage of development, and has differential effects on the cranial and caudal parts of the ducts, and even contralaterally; after castration the Wolffian ducts on the right side require more androgens for stimulating the differentiation than those on the left side. In intersexes of man rudimentary testes are more frequent on the right side and rudimentary ovaries on the left. In the normal human foetus the right gonads tend to be larger and to have more protein and DNA than those on the left.

In another abnormality in males with XY constitution, the testes are atrophic. This phenomenon is called testicular feminization and is due to an insufficient quantity of the enzyme α-reductase, which converts the testosterone into its active form of dehydrotestosterone. In some strains of mice testicular feminization is due to a deficiency of androgen receptors in XY mice. These animals have small testes, but no other internal reproductive organs derived from Wolffian ducts, and have external genitalia of female type.

It is significant that foetal hormone production by the male affects the sexual differentiation of the twin female of some species when the placental vessels anastomose, but that there is no converse effect of the female on the male twin. This fact is in agreement with the observation that, unlike the testis, the foetal ovary does not produce steroid hormones in any quantity. During pregnancy oestrogens are secreted by the mother, in amounts varying with the stage of pregnancy, and by the foeto-placental unit which consists of the placenta and the foetal adrenals. Whether the amounts of oestrogens produced by the unit differ in the male and female conceptus has not been established. It is feasible that the maternal and foeto-placental output is sufficient to inhibit an unnecessary secretory activity in the foetal ovary, and make it necessary for the foetal testis to produce androgens to counteract a possible oestrogenic interference with the differentiation of the male genital tract. Whether this secretion at early foetal stages is mediated via the centres in the pituitary and hypothalamus is not known.

The balance of chromosomal and hormonal factors in sex determination is different for males and females; early secretory activity of the foetal testis is essential for development, but not required from the ovary. The female is susceptible to hormonal influences at these periods of development, as is clearly shown by the appearance of freemartins and of intersexes. Steroid hormones are not wholly responsible for such phenomena, since administration of androgens does not have the same effects as testicular implants, which are attributed to the additional action of a factor X. The steroid hormones are usually bound to specific proteins for controlled release at the right time and in the right location, and are effective only if the target organs have the specific receptors. The inadequacy of the receptors results in testicular feminization in mice with a faulty genetic constitution (Tfm), and insufficiency of an enzyme accounts for a similar phenomenon in man. In the adult the production and release of the gonadal steroid hormones is controlled by tropic hormones of the pituitary, and their release is governed by releasing factors in the hypothalamic nuclei. Final determination of sex involves the differentiation of these centres in the brain, and this process takes place during the first weeks following birth in mice and rats, as will be discussed after the migration of the primordial germ cells and their fate in the gonads have been described.

The number of gonocytes in the testis of a mouse increases from about 100 to 5000 between the 10th and 14th day of gestation, partly by the arrival of further cells from the yolk-sac, partly, and probably mainly, by mitosis. Cellular multiplication stops after 14 days in the mouse and 18 days in the rat. Of those present at parturition 50% die in the first postnatal week. The remaining gonocytes start to divide again about 8 to 10 days after birth in the mouse and 11 to 14 days in the rat. The process of spermatogenesis starts with puberty, usually at 4 weeks in the mouse and 10 weeks in the rat. In the rat the number of interstitial cells starts to decline at birth, reaches a low level between the 6th and 30th postnatal day and subsequently increases, and the cells enlarge. Androgen production also declines from the high plateau reached on the 18th day of gestation to a very low level at between the 15th and 30th day post partum, when it rises with the increasing number, size and activity of the interstitial cells.

The number of gonocytes in the human foetal ovary reaches its peak of 7×10^6 at 5 months, declines subsequently by degeneration of cells to about 1×10^6 at birth and to 3×10^5 at about 7 years. The total number of ova available for the whole reproductive period between puberty and menopause is about 500. In women, monkeys, bovines and rodents all those present at birth are in the stage of prophase of the first meiotic division, and do not multiply by mitosis, while in rabbits, cats and hamsters, this stage is reached shortly after birth; in most mammals the production of ova therefore seems to end in the perinatal period. The interstitial tissue of the ovary comprises the cells of the follicles which form around the ova and the theca cells which surround the follicles. They become active at puberty under the influence of pituitary hormones whose release is regulated by the hypothalamus. These centres impose the cyclical activity evident in the oestrous and menstrual periods due to successive phases of oestrogen or progesterone production by the interstitial tissue.

In both sexes the number of primordial germ cells at first increases rapidly, but towards the end of the foetal period declines to about 50% in the male and to a mere 0·05% of those present at birth in females. While the number of spermatogonia increases by mitosis after puberty and throughout life, that of oogonia decreases steadily in all mammals and reaches its nadir at the menopause of women. In males the secretory functions of the gonad diminish after birth and are resumed at puberty, while spermatogenesis is dormant throughout the foetal and postnatal period up to puberty. The secretory functions of the female gonad are in abeyance during the foetal and postnatal period up to puberty, while the oogonia multiply rapidly throughout the foetal period, and on reaching the prophase of meiosis at about birth cease to multiply. The two aspects of gonadal function, the production of germ cells and the endocrine activities of the interstitial tissues, have thus a divergent origin, and the development of the latter differs with sex; in the male the secretory functions of the foetal gonad are well developed and play an essential role in the differentiation of sexual characteristics, while the spermatogonia and their supporting cells (Sertoli cells) are inactive and undifferentiated. In the female foetal gonad the interstitial tissue is inactive and does not play an important part in the sexual differentiation of the individual, while the oogonia are very active and multiply and differentiate to an advanced stage. The two gonadal functions are coordinated at puberty in both sexes.

Some of the gonadal differentiations in the foetal period depend on secretions by the foeto-placental unit in which the placenta and the foetal adrenal complement each other's activities. Precursors of various steroids and of peptide and protein hormones arrive via the maternal circulation, and are converted at various stages by enzymes of either the placental tissue or the adrenals. Oestrogens are produced in this manner from androgens which may be secreted by the foetal adrenals. Human chorionic gonadotropins (HCG) are glycoproteins and formed in the placenta on the 9th to 10th day of pregnancy, and then increase in amounts which reach and maintain a peak value at about 10 to 16 weeks. They subsequently decline quantitatively up to the end of gestation. It is not known whether they differ in amounts with the sex of the foetus, but they are thought to be involved in the descent of the testis from the kidney region to the scrotum. If the testis remains inside the body cavity (cryptorchism), it atrophies and fails to produce spermatozoa. Treatment of boys with HCG can correct this developmental error. Oestrogens are secreted largely in the foetal zone of the adrenals, which is formed by an enlarged outer zone of the adrenal cortex and which regresses after birth. If the foetus dies, the output of oestrogens falls sharply, though the HCG level may be maintained for some weeks after the death of the foetus. The foeto-placental secretions are largely responsible for the maintenance of pregnancy, affect the maternal as well as foetal tissues, and are involved in coordinating the events in both at the termination of pregnancy and at parturition. The role of regulatory factors in the foetal brain in these processes, and earlier in directing the activities of the adrenals and other endocrine glands, is probable, but not fully understood. There is no evidence for a sexual dimorphism in their functions at this stage.

5. Neuroendocrine factors in sex determination

The chromosomes of the fertilized ovum form the foundation of the subsequent sexual differentiation of the individual, but the development of the sexual characters is completed only after puberty, when a variety of somatic and psychological changes occur apart from the achievement of sexual maturity. These are influenced by the gonadal differentiation, particularly of the interstitial tissues which at first respond to stimuli from the foeto-placental unit and after birth to those from the centres in the brain. The differentiation of these functions in the pituitary and the hypothalamus may not be completed at the time of parturition, and may become operative only at the time of puberty. This differentiation varies in its timing with the duration of the period of gestation and with the degree of maturity of the foetus at birth. In larger mammals with long gestation periods, such as men and monkeys, the differentiation of the centres may be completed in the later stages of pregnancy. In mice and rats this process takes place in the first postnatal weeks and can be influenced during a critical period easily in the first 10 days and with greater difficulty during the following 20 days.

The centres have two layers which collaborate closely and are linked by a portal system of blood vessels. The capillary network of the hypothalamus discharges its contents into veins which enter the pituitary, and form a network of capillaries, so that the factors produced by the hypothalamic cells reach their target cells in the hypophysis without being diluted in the general circulation. A similar system operates in the portal circulation of the liver. The venous blood from the intestines is brought first to the liver by the branches of the portal vein, which forms a capillary network around the liver cells which metabolize proteins, carbohydrates and fat, and discharge the metabolites into the capillaries of the hepatic and cava veins, to enter the general circulation, and later reach other parts of the body with the arterial blood.

The hypothalamic nuclei are formed by groups of ganglion cells which produce specific releasing or inhibiting factors for target cells in the pituitary. They differ from other ganglion cells, which produce transmitter substances carried via the nerve fibres to a synapse, both in the type of their product and the transmission via the circulation instead of by nervous connections. They are described as neuroendocrine cells, indicating that they are ganglion cells, but that like other endocrines they release their secretions into the circulation. There are also some nervous fibres linking the hypothalamus and the pituitary, particularly its posterior region called the neurohypophysis.

These neuroendocrine cells are grouped into nuclei within the hypothalamic region of the brain, and produce specific agents which signal cells in the pituitary to release or withhold their specific tropic* hormones which activate the thyroid (thyrotropic cells and hormones), the gonads (gonadotropic cells and hormones), the adrenal cortex (corticotropic cells and hormones), the growth hormone, which after conversion in the liver affects almost all tissues, and other hormones. There are also agents produced in the hypothalamus which act as inhibitors for the release of prolactin, a hormone concerned with the secretion of milk in the breast and produced in the pituitary. Whether the factors 'release' or 'inhibit' is of importance in the feed-back relations to the concentration of the specific hormones in the blood reaching the hypothalamus: a low concentration of oestrogens in the blood around the hypothalamic nuclei will signal the need for the release of the specific hormone from the pituitary (follicle-stimulating hormone, FSH) to stimulate the oestrogen synthesis and release by ovarian follicular cells. In the case of prolactin a high blood level of this tropic hormone will act as a signal for the release of inhibiting factors from the hypothalamus to stop the production of prolactin in the pituitary, and thus later of milk in the mammary glands.

The connections between the hypothalamic nuclei and the posterior pituitary are both nervous and vascular, and are concerned with the release of an antidiuretic hormone acting on the kidney, and with that of oxytocin, which affects the extrusion of milk from the breasts and the activity of the uterus during parturition. We are here mainly concerned with the role of those structures in the hypothalamus and the pituitary which affect the sexual development directly, i.e., the gonadotropic complex, to which must be added the influence of the pineal body, which is situated at the posterior end of the roof of the third ventricle of the brain, and is known to affect the onset of puberty and to be involved in the transmission of the effects of the photoperiod. This implies that perceptions of light from the eyes, and

---

*The correct spelling is 'tropic', but the wrong form 'trophic' is commonly used in scientific publications.

probably also the skin, are referred to the pineal body, and that either in direct response to it or more likely in response to the activity of hypothalamic cells which stimulate the pineal body, a substance called melatonin is produced which interferes with the secretion of gonadotropins. This point is mentioned to emphasize that the hypothalamic nuclei are in contact with other parts of the brain concerned with the perception of and reaction to sensory and other stimuli. They are thus capable of regulating the sexual development, and functions in response to environmental influences as well as to the activities of endogenous peripheral endocrines and other organs.

While the hypothalamus has a central role in coordinating the input and output of the various interacting organs and structures involved in the development of the sexual functions, changes at any point in the chain of processes have repercussions on the other members of the system, encompassing the receptor sites at the ultimate target cells, the peripheral endocrines, the pituitary, the pineal body and the hypothalamus itself. Interruptions may occur at different points with similar effects, which may make it very difficult to pin down the real cause of any abnormality. These difficulties are increased, since failures at some points may elicit compensatory reactions at others. Thus failure of the gonads to develop may be due to chromosomal factors, or to local abnormalities such as enzyme deficiency, or to failure of receptors, or to local endocrine effects regulated by tropic hormones and their regulation by higher centres. Even for identifiable causation by chromosomes (e.g. XO, XXY, XYY) or by faulty genes which interfere with the sexual development, it is by no means clear at what stage or in which process they exert their effect.

The interactions between the gonads and the higher centres have been investigated experimentally for at least 50 years, mainly by transplantation experiments in which gonadectomy has been succeeded by grafts of the gonads from the opposite sex; ovaries have been transplanted into castrate males and testes into spayed females. The results are assessed by changes in the appearance of the external genitalia, and by additional grafting experiments of, for instance, the ovaries into the anterior chambers of the eye in castrates or spayed animals to detect ovarian cycles by the periodic growth of follicles. These experiments were refined and amplified by hypophysectomies followed by grafts of gonads-plus-pituitary, and ultimately of both in combination with the hypothalamic region. By these means the fundamental facts of the feedback relationships between the gonads, pituitary and hypothalamus have been established. In the last 20 years, purified hormones and extracts of the relevant tissues have simplified the experimental procedures, and contributed further detailed information of a quantitative nature. Most of these experiments are performed in mice and rats shortly after their birth. In these species, which have a short gestation period, the centres in the brain are not yet sexually differentiated, while they differentiate during the pregnancy in guinea-pigs and rhesus monkeys. The experimentation in the intra-uterine foetus may endanger the successful end of the pregnancy, and is more difficult technically than similar procedures applied to the newly-born offspring. In mice and rats, there is a critical phase for the differentiation of the centres lasting for the first 10 postnatal days, during which the development of the hypothalamus and the pituitary is easily affected by the injection of androgens and oestrogens into males and females respectively. Such treatments influence the onset of puberty, the development of the external genitalia, the cyclical pattern of sexual activities, and the sexual behaviour of the adults, as well as non-sexual characters such as the growth rate, the proportions of the body, the differential growth of the skeleton and muscles, the amount and location of fat deposits, the size of the kidneys and adrenals, and a variety of metabolic and physiological functions. The results vary in detail with species and strain of animal, with the preparation used, and with the dosage and timing of its application.

During the critical period the type of sexual differentiation is altered by male and female hormones (gonadotropins, progesterone, oestrogen, androgen), by some specific procedures (hypothalamic lesions, parabiosis of gonadectomized with normal animals), as well as by the administration of corticosteroids, cholesterol, antiandrogens and by continuous illumination, continuous darkness or exposure to continuous noise. The similarity of the effects of these diverse agents, particularly of those mediated by the sense organs, suggests that their target is the central region rather than the peripheral endocrines. This is confirmed by additional transplantation experiments, by hypophysectomy or ablation of the hypothalamus. The effects are most marked when the experiments are performed during the first 10 days of life; they require greater dosages of the various stimuli later on and subsequently merely equal the effects of gonadectomy in adults. It has long been known that castration to preserve the pitch of the boyish voice in professional singers by preventing the enlargement of the larynx is more effective before puberty than later on.

Female mice and rats have an oestrous cycle of ovarian activity and changes in the genital tract. In rats the cycle lasts precisely four or precisely five days, but is not as rigidly timed in mice. The ovarian activity culminates in the rupture of mature follicles and ovulation. This stage is preceded by the 'follicular' phase in which the small follicles enlarge and mature and followed by the 'luteal' phase when a corpus luteum is formed by the remaining follicular and perifollicular cells. This ovarian cycle is reflected in changes in the uterus and the vagina and governed by two pituitary gonadotropins: the follicle-stimulating hormone (FSH) and the luteinizing hormone (LH), which are regulated by specific releasing factors of the hypothalamus. The object of these changes is to prepare the female for successful mating at oestrus, and subsequently for the implantation of the fertilized ova in the uterus. In the preparatory follicular phase, secretion of oestrogen is stimulated by FSH released from the pituitary at the command of the releasing hormones of the hypothalamus, which are activated by a low concentration of oestrogen in the blood around the hypothalamus. FSH stimulates the follicle cells of the ovary to multiply and to secrete oestrogens, which in turn promote the growth of the inner layers of the uterus, and of the surface layers of the vagina, and the growth and secretory activity of the associated glandular structures; they also increase the vascular supply and alter the permeability of the vessels. The behaviour of the female is affected, as is shown by the characteristic arching of the back (lordosis) and her readiness to accept the male at the height of the oestrus. This stage is followed by the bursting of the large follicles and ovulation, and by the discharge of the ova into the tubes where they may be fertilized by the spermatozoa.

The high oestrogen level of the blood reaching the hypothalamus signals a halt in the releasing factors for FSH, and causes a decrease in the ovarian production of oestrogens. Meanwhile, a low level of LH and consequently of progesterone in the blood is activating the hypothalamus to release the releasing hormone for LH, which turns on the gonadotropins of the pituitary, and LH stimulates the follicular and perifollicular cells to produce progesterones after ovulation. These ovarian hormones act on the innermost epithelial and glandular layers of the uterus (the endometrium), causing them to increase in width and secretory activity and thus to prepare for the implantation of the conceptus. In the vagina, the horny surface layer is replaced under their influence by a mucin-producing sheet.

The stage of the cycle is ascertained by the microscopic examination of cells in vaginal smears scraped or pipetted off the surface of the vagina. At the height of the oestrus almost all cells are cornified and large, with small pycnotic nuclei or none, a sort of dandruff. In the follicular phase many nucleated larger cells are seen, either alone or mixed up with some cornified cells when approaching oestrus. After oestrus, a few cornified cells and many nucleated cells appear, together with great numbers of white blood cells. If there is no implantation of a conceptus, the inner layers of the uterus and vagina regress as the output of progesterone is halted and that of oestrogens is resumed. If the mating has been successful and fertilized ova are implanted, the cycle is stopped and a corpus luteum of pregnancy is formed.

If animals are treated in the critical neonatal period with androgens or oestrogens or some of the other substances or procedures, the cycle will be arrested, either at oestrus with persisting cornified layers in the vagina or at anoestrus, when the uterus as well as the vagina will appear atrophic. The effect, or at least its symptoms, are delayed until puberty and are not immediately apparent, thus pointing to an alteration in the differentiation of the centres.

The female cycle can be abolished by interference at any of the major links in the feed-back system, and this makes the localization of a biological clock for the process rather difficult. Removal of the ovary, and thus of its hormones, causes atrophy in the peripheral target organs such as the uterus, vagina and breast, and at the same time leads to hypertrophy (enlargement and increased activity) of the gonadotropins in the pituitary and changes in the hypothalamus, which may result in tumour formation, because the persistently low levels of oestrogens and progesterones in the blood stimulate the activity of the hypothalamus and, through it, that of the hypophysis. Removal of the pituitary causes atrophy of the ovary and secondary atrophy of the uterus and vagina and abolishes the cycle, while stimulating hyperactivity in the hypothalamus. If the ovaries are excised and grafted onto the spleen, the ovarian secretions will reach the liver via the splenic vein and are metabolized in the liver. Thus their level in the general circulation will be reduced, and the hypothalamus will be excited to stimulate the gonadotropins of the pituitary by releasing hormones; these act on the ectopic (displaced) ovary which responds with greatly increased activity which results in the appearance of ovarian tumours. If the excised ovary is placed under the skin or into other situations without prior drainage into the liver, the hormone level in the blood fluctuates regularly and the ovary remains normal.

The role of the hypothalamus in the cyclical events can be demonstrated by destroying it with an electric current or surgically, but more elegantly by a series of transplantation experiments. If the pituitary of a female is removed and either a male or a female pituitary placed close to the hypothalamus, normal cycles are resumed. In the reverse experiment, a male is castrated, hypophysectomized, and grafted with an uterus and ovary, and has a female pituitary placed close to the hypothalamus. The female cycle is not restored, thus showing that the hypothalamic control supersedes the presence of the female pituitary and ovary. At this stage, the sexual differentiation of the hypothalamus is complete and apparently permanently determined, while the pituitary merely follows the commands of the hypothalamus and changes accordingly from a male to a female function and vice versa. This is also shown in female rats androgenized when five days old, and consequently sterile and without cyclical activity. If the anterior pituitaries from such animals are placed close to the hypothalamus of other females which after removal of their own pituitary have no cycles and atrophic ovaries, the pituitaries taken from the androgenized females restore normal cycles in their hosts under the guidance of the host's hypothalamus. This experiment also proves that the treatment of mice and rats in the critical neonatal phase with various substances and procedures affects primarily the hypothalamic region, which, once differentiated, persists with the characters acquired, while the pituitary and the peripheral endocrines are not permanently determined as regards the pattern of sexual endocrine activity. This is finally established by an experiment in which the hypothalamus of a castrate male is destroyed and replaced by a female hypothalamus close to the pituitary, and is also provided with an ovary and uterus. Under these conditions, cyclical activity in the ovarian secretions is restored. The sexual differentiation of the hypothalamic centres in the critical postnatal phase in rodents, and during pregnancy in other species, is thus an essential step in the determination of sex.

Treatment of male mice and rats with oestrogens during the first ten postnatal days inhibits the secretion of androgens and growth of the testis, impairs or prevents spermatogenesis, and interferes with the development of the external genitalia and the accessory glands. The effects resemble those of castration, and do not change the differentiation of the hypothalmus, but treatment with antiandrogens does. The female type of hypothalamic differentiation induced in treated males can be demonstrated by periodic follicular activity in implanted ovaries, and cycles in transplants of uteri and vaginae. Treatment with antiandrogens also inhibits the development of the prostate, of the seminal vesicles and of the penis with its bone, and induces hypospadias by incomplete closure of the urethra. It stimulates the development of the mammary glands and the formation of nipples, which are normally absent, and of lactation if prolactin is given in addition. The behaviour of the antiandrogenized males is of female type, and they are accepted as females by other rats. The difference between effects of the administration of oestrogens and of antiandrogens underlines the conclusion that even small amounts of androgens in oestrogenized males are sufficient to suppress the female differentiation of the hypothalamic centres, though they are not sufficient to maintain the normal development of the testes and internal as well as external genital structures. Effects comparable to the androgenization of females can be achieved by antiandrogens, but not by oestrogens.

The critical period for the differentiation of the neuroendocrine centres occurs in the neonatal period of rodents and during the intra-uterine period in guinea-pigs, dogs and rhesus monkeys, and at that time similar results are produced by treatment with androgens and antiandrogens. As the secretion of androgens by the foetal testis appears to suppress the female differentiation in mammals, and since its formation is initiated by the Y-chromosome of the embryo, it might be argued that this chromosome is ultimately responsible for the entire process of sex determination. This conclusion is not justified, since no distinctive genes on the Y-chromosome are known which might account for the successive steps in the sexual development, which are governed by gonadal and neuroendocrine hormones. These in turn are susceptible to such external influences as light, noise, temperature and nutrition, and interactions with other endogenous conditions. The examples of the appearance of freemartins in the presence of a male twin and placental anastomoses, and of the role of androgens and antiandrogens at later foetal or postnatal stages, show that at these periods the sexual development is not predetermined by the presence of a Y-chromosome, but that the whole system is still malleable. The process of sex determination consists of a series of steps following the initial stage of chromosomal action. At first the gonadal hormones, which are steroids, play a prominent part, and they continue to exert their influence after the foetal period in the shaping of the body and that of its structures and functions. Whether they are produced initially quite independently of directions from the central nervous system, is not clear, but they are certainly subject to the gonadotropins secreted by the foeto-placental unit. The neuroendocrine factors are polypeptides, which become operative in the late foetal period of a species with a long gestation period and in the postnatal period in those where it is short.

The development of the feed-back regulations between the gonads and the higher centres in the brain does not end with birth, and additional endocrine (thyroid, adrenal, thymus) and possibly paracrine factors come into play. The latter are of two main types;

prostaglandins are derived from fatty acids and are produced by most tissues, and the APUD system (named from the function of amine precursor uptake and decarboxylase) are peptides produced in the brain and intestine. Both types of substance act mainly on adjacent cells and tissues, in contrast with the transport of hormones and neuroendocrine factors in the blood vessels or nerve fibres. Whether the paracrine agents vary with sex is not yet established, but it is not unlikely, since the prostaglandins are involved in the functions of the uterus and prostate, and the APUD factors in those of the hypothalamus, pineal body and the adrenal cortex. These substances have been discovered only recently, and the investigation of their functions is still at an early stage. An agent acting at a short distance only and secreted by the foetal testis has been mentioned as factor X or medullarin. Neither its nature nor the group of paracrines to which it may belong is known.

6. Sexual maturation and decline

The steroid sex hormones not only act in the determination of sex, but influence the rate of growth and differentiation of somatic tissues, the configuration of the body, the deposition of fat, the metabolism, and a variety of functions which will be described in later chapters. Though these hormones are labelled 'male' and 'female', it should be remembered that they are present in both sexes, albeit in different proportions of roughly 1:10 in the adult phase, though the balance shifts at different periods of life. This chapter deals with the question how puberty and breeding seasons are induced, and with the problem of the menopause.

Gonadal endocrine activity in most species of mammals subsides after birth, with the cessation of the foeto-placental activity by the discarding of the placenta as the afterbirth and the regression of the foetal zone of the adrenal cortex. This affects the testis particularly, as the endocrine activity of the foetal ovary is minimal. As mentioned above, the interstitial cells of the testis are reduced in number and activity after parturition and so the amount of steroids produced falls also. This state lasts until puberty, though some steroids are secreted by the adrenals in either sex, as the cortex of the permanent adrenal differentiates.

Puberty is not an instantaneous event, though the first menstrual bleeding in girls may suggest a sudden change. In man as in many mammals, females reach puberty before the males, often one or more years earlier. This is true for apes and monkeys, for elephants, deer, pigs, voles and many other species. Exceptionally in the bat *Megaderma lyra*, males become sexually mature earlier than females. There is apparently no difference between sexes in the time at which puberty is reached in mice, rats, sheep and cattle. It is difficult, however, to determine the onset of puberty with exactitude. The starting of sexual cycles in females is a fairly obvious phenomenon, but spermatogenesis is more difficult to date, as its onset is gradual and it is a slow process. In addition to the activities of the reproductive organs, various somatic and psychological or behavioural developments characterize the pubertal period, and these are continuous and slow processes without clear dividing lines. They occur almost imperceptibly over months or years depending on the species.

In girls, the menarche, the onset of menstruation, has during the last 100 years advanced from about 17 years on the average to 14 years or less in the Western world. The reason for this acceleration in maturation is not at all clear. It may be related to improvements in nutrition and the consequent increased rate of growth, which is manifested also in the greater average height. It may be due to changes in the type of food available, which may affect the endocrine system. External stimuli such as light, temperature, sensory stimulations and psychological factors are known to accelerate or inhibit the onset of puberty in other mammals. Thus in mice the presence of a single male during the weaning period hastens the onset of puberty in females. This effect may be due to the excretion of pheromones, chemical messengers, excreted by the male and exciting the olfactory system and through it the hypothalamic centres of the suckling females. Such factors are present in the urine of adult males and females, and act both as sexual attractants between the sexes and as repellants for the same sex, as for instance by the marking of territories by males of various species through their urine. The marking and smelling of lamp posts and trees by dogs is a familiar example. Oddly enough, blindness in girls advances the menarche, while continuous lighting speeds up the sexual maturation of male and female rats. Lowering of the temperature inhibits growth of mice and delays the onset of puberty. Too little food delays puberty in rats, while too much food does so in piglets. Intake of protein has an influence on sexual maturation, and deficiency of vitamin A delays it in female rats and of vitamin E in male rats. The red kangaroo (*Megaleia rufa*) reaches puberty at 27 months in regions with abundant food, but not until 35 months when food is scarce. Thus a number of environmental factors are implicated in the timing of puberty, and some of these could be involved in the acceleration of the menarche in girls during the last century. It is not evident, however, whether these factors act on the hypothalamic centres, the pituitary and pineal body, or affect the sensitivity of the gonads to gonadotropins. Obviously the final result depends on the coordination of all these activities and is achieved only slowly. This is shown by the initial occurrence of anovulatory cycles in girls, and of abortive prepubertal waves of follicular development in mice and guinea-pigs. There is the further possibility of a change in the metabolism of oestrogenic steroids in the liver, which has been observed in rats at the time of puberty. The implication of the pineal body is based on the acceleration of the onset of puberty in rats by pinealectomy, and its delay by the administration of extracts of the pineal gland or of its specific hormone, melatonin. In girls, lesions in the pineal body advance the menarche. In hamsters, pinealectomy prevents the testicular atrophy that occurs if the animals are kept in the dark. These findings link the sensitivity to light in sexual maturation to that of the pineal gland as probably an intermediary between the sensory receptors and the hypothalamic-pituitary complex, which regulates the gonadal functions and their onset. It is not yet possible to assign a determining role, either to exogenous stimuli or to endogenous factors related to growth, for the triggering of the activity of the centres and the gonads at puberty.

Secondary to the onset of gonadal secretion at puberty are the differentiation and growth of the internal and external genitalia. In boys, the testes and the penis with its corpora cavernosa enlarge, the prostate and seminal vesicle and accessory glands in the genital and axillary region increase in size and start to secrete, the sexual hairs in the pubic, axillary, chest and facial regions replace the vellus hair coat. All these developments are slow processes extending over a span of months and even years. In girls, differentiation and increase in volume of the uterus, the vagina, their glands, of the labia and clitoris in the vulva, and of the breasts, are comparable features of sexual maturation. They again are slow processes, taking months or even years. In animals, similar changes take place, though varying in detail and timing. In female rodents, the opening of the vagina is considered as the criterion for the onset of puberty and the effects of perinatal treatment with hormones and other substances become manifest at this stage; if animals are treated with androgens instead of oestrogens the opening is delayed or the vagina may not open at all, while the clitoris enlarges and may form a small bone like that of the penis, the uterus and vagina do not develop, while the breasts may be larger than usual. The left and right sides of the clitoris may fail to unite, and thus give rise to a hypospadias. The ovaries of androgenized females are atrophic and devoid of corpora lutea, and oestrous cycles, if occurring at all, are very irregular.

In males, perinatal treatment with oestrogens, and more so with antiandrogens, inhibits testicular growth and spermatogenesis, the enlargement and secretory activities of the accessory glands, and that of the penis and the bone and corpora cavernosa in it.

The oestrous cycles in females determine the preparation for mating, the release of ova for fertilization, and the preparation of the uterus for the implantation of the fertilized ovum. Timing is essential, since both the ovum and the sperm have only a short period in which they are capable of fertilization, and it is up to the female to ensure that the mating is effected at the right time. This involves behavioural signs as well as the release of pheromones, often in the urine and vaginal excretions. The first part of the cycle is the follicular phase, in which release of FSH stimulates the production of oestrogens in the follicles and thus causes the enlargement of the uterus and vagina, and changes in the inner lining of both organs. These changes, which accompany the growth of the follicles, culminate in the discharge of ova into the Fallopian tubes after the bursting of the mature follicles. This phase of oestrus (heat) is signalled also by a curvature of the spine so that it is convex ventrally (lordosis) showing readiness to accept the male. Ovulation is preceded by a rise in LH as well as FSH and is followed by a decrease in FSH and oestrogen production, and a rise in LH and production of progesterones by the cells of the burst follicle and the cells surrounding it. During this luteal phase of the cycle, the uterine glands enlarge further and the inner layers are made ready for placentation, while the cornified layers of the vaginal lining, formed under the influence of the oestrogens, are shed and are replaced by a mucin-secreting layer. Fertilization must occur within hours after ovulation, and is usually accomplished in the tubes. Implantation of the conceptus stops the cycle, and the foeto-placental unit develops to play its part in the maintenance of the pregnancy. If there is no implantation, the cycle resumes; the high level of circulating progesterone signals to the hypothalamus that the release of LH should stop, and the low level of oestrogens that the releasing factors should stimulate the pituitary to produce FSH and stimulate the follicular growth and oestrogen production in the ovary. The corpus luteum in the ovary and the inner layers of the uterus regress, and multiplication of cellular layers of the vaginal lining begins again. The female will accept the male only during the short phase of oestrus. The ova have reached the prophase of the first meiosis at or before birth, and need only complete the first meiotic division at ovulation and the second meiotic division following the entry of the spermatozoon. The oestrous cycles are thus concerned less with the maturation of the ovum than with the preparations for its release by the surrounding cells and the reception of the fertilized ovum in the uterus. The FSH and LH influence mainly the follicular and surrounding cells, and their secretions of oestrogens and progesterones affect the internal and external genital organs; it is therefore these structures which are changed, rather than the gonocytes at puberty in the female.

In males, the testes are undeveloped as regards the gonocytes and their specific supporting cells (Sertoli cells) at birth and before puberty, while the interstitial tissue has declined in volume after its foetal activity. Pubertal changes in the male gonad thus have to affect both the endocrine tissue and the seminiferous tubules concerned with spermatogenesis. The testicular weight increases by the enlargement of the seminiferous tubules, where spermatogenesis takes place, and by augmenting the number and size of the interstitial cells as well as the supporting structures of blood vessels and connective tissue. The interstitial tissue is stimulated to the production of testosterone by the excretion of the interstitial-cell-stimulating hormone (ICSH) in the pituitary, which is comparable to, if not identical with, the LH in females, and frequently described as LH. The seminiferous tubules form lumina, and Sertoli cells differentiate, while the spermatogonia start to proliferate and to differentiate into the various stages leading to the appearance of spermatozoa. This process is stimulated by the pituitary FSH, which in the presence of testosterone activates the initial spermatogonia. Spermatogenesis proceeds along the tubules in waves as a continuous process, starting successively in adjacent cells. Only the initial step seems to require the action of FSH, and the subsequent differentiation until the stage of the first meiotic division does not require any further stimulation. The Sertoli cells differentiate and serve as nutrient cells for the spermatids with which they have close connections. FSH and LH in the male, as in the female, are controlled by hypothalamic releasing factors, and there are feed-back relations between testosterone and LH, and between FSH and a substance called inhibin in the seminiferous tubules, which indicates the level of production of spermatozoa. The output of LH and the production of testosterone have a diurnal rhythm with a low point in man at between 1200 and 1800h for LH and slightly later for testosterone. There is no similar diurnal rhythm in the output of FSH in males and no cycle of days or weeks in the activity of male gonads, as there is in the female.

The menstrual cycle of women differs from the oestrous cycle in non-primate mammals in the reaction of the inner layers of the uterus, which are partly shed at the end of the luteal phase. Three stages can be distinguished; (a) the proliferative or oestrogenic period, which ends with ovulation on about the 14th day (b) the luteal or postovulatory period, and (c) menstruation, the shedding of the endometrial structures and bleeding. The last phase is due to the withdrawal of oestrogen. The follicular activity is controlled by FSH, the formation of a corpus luteum by LH as in other mammals with an oestrous cycle, and there are very similar feed-back regulations with the hypothalamus and the pituitary. As in rodents, pregnancy arrests the menstrual cycle.

Sexual maturation and decline

The duration and frequency of the oestrous cycles vary with species. They may occur throughout the year or during the breeding season in series in a polyoestrous species, or may be isolated events in a monoestrous species. Rats and mice breed throughout the year and are polyoestrous, with each cycle lasting 4 or 5 days. Even here the duration of the cycle varies with the photoperiod, nutrition, temperature and a number of environmental factors such as the density of the population within a cage, and the presence of other males or females. Cycles tend to be regular in female mice if groups of 4 are confined in a cage, but irregular if their number is increased to 30 or lowered to 1. The introduction of a single male into a cage with females induces simultaneous cycles. The presence of a strange male or even of his urine blocks the progress of pregnancy by inhibiting the functions of the corpora lutea. This effect is probably mediated by pheromones. In guinea-pigs the cycles last about 15 days, in the pig and cow about 3 weeks, in monkeys and women 4 weeks, in the chimpanzee 5 weeks. These species breed throughout the year and are polyoestrous. Seasonal breeders may be polyoestrous or monoestrous; sheep, goats, horses and cats are polyoestrous with cycles of 3 to 7 weeks, while the dog and mole are monoestrous with only 1 or 2 oestrous cycles throughout the year.

The onset of the breeding season varies with species; in sheep, it coincides with the shortest days of the year, while the lengthening of the day stimulates pituitary secretion in horses and donkeys, and the breeding season reaches its peak in April. Vixen come into oestrus in February. Camels cease to breed during the hottest part of the year, brown rats in Alaska during the coldest. The internal clock of the oestrous cycles is thus subject to a number of environmental influences which are probably mediated via the pineal body, the pituitary and the hypothalamus, and affect females predominantly. Billy goats, rams and dogs are capable of mating throughout the year, if they come into contact with females in heat. In some species, male fertility declines at the end of the breeding season, and this is evident in the loss of weight and cessation of activity in the testes. In foxes, the testis starts to increase in weight in December, reaches a peak in February coinciding with the oestrus of vixen, and decreases again in April. Some mammals withdraw the testis from the scrotum into the body cavity outside the breeding season, as for instance the Alaskan rat. The scrotum dissipates heat, and its temperature is about 2 °C lower than the internal body temperature of man, and a rise in temperature impairs spermatogenesis. Dramatic somatic and behavioural changes occur in the red deer stag. During the breeding season, which reaches its peak in September, the dominant male is very aggressive, and collects several hinds into a harem, which he defends from other males. From October onwards he becomes less aggressive and associates peacably with his rivals, as his testicular activity declines during the winter and spring. The old antlers are cast in April, and the gonads resume their activity in July, when new antlers are formed. His social behaviour then changes, and he fights with other males to assert his dominance during the breeding season.

The processes and environmental stimuli leading to the onset of puberty and of the breeding season are similar. Both involve environmental conditions of a great variety as stimulants of the brain centres via the senses and, transmitted to the gonads, influence females more than males. This may be due to some extent to the fact that the cyclical activity of females is more easily upset than the continuous operation in males, which is subject only to minor diurnal variations. Endogenous factors arresting the female cycles are pregnancy and lactation which act on the regulatory centres. In pregnancy it is the secretion of chorionic gonadotropins at first by the trophoblast, subsequently by the corpus luteum of pregnancy, and later by the foeto-placental unit. In some species the corpus luteum is not necessary for the maintenance of the later stages of pregnancy. Lactation is regulated by the hypothalamus and the pituitary by prolactin, which is released when the inhibiting factor of the hypothalamus ceases to act, and this starts in the later stages of pregnancy. The exact chain of processes leading to the onset of puberty and of breeding seasons is not known. Nor is it clear why animals such as the dog and mole are monoestrous, while some of the other seasonal breeders are polyoestrous, nor why shortening of the day stimulates in some species and lengthening in others. Also unknown is why blindness advances the menarche of girls and continuous lighting the puberty of rats.

The difference with sex in the maturity of gonocytes reached at the onset of puberty is reflected in the time required for the maturation of ovarian follicles and for spermatogenesis. The ova have reached the meiotic prophase at birth, and the second meiotic division is completed only after fertilization. The process of maturation of the oocyte and formation of follicles is restricted to the first half of the oestrous or menstrual cycle and thus to about 2 days in rodents and 2 weeks in women. The stages of spermatogenesis from the initial spermatogonia, usually called 'A', to the formation of spermatozoa include both meiotic divisions and take about 74 days in man, 60 days in bulls, 52 days in rabbits, 49 days in rams and rats, 45 days in mice and 35 days in boars. Though spermatogenesis is a continuous operation starting in successive waves of spermatogonia, the differentiation is a lengthy and very involved process in comparison with the ovarian cycle which uses up the follicles and ova available after puberty.

In the sexually mature female, the gonadotropic hormones affect the cells surrounding the ovum rather than the ova themselves, while in the male, they act on the interstitial tissue, the Sertoli cells and the gonocytes themselves; the proliferation of spermatogonia A is stimulated by FSH in the presence of testosterone, which is regulated by LH. The subsequent stages of mitotic multiplication of spermatogonia do not require the presence of hormones up to and including the pachytene stage of the first meiotic division, though the completion of this process requires testosterone. These conclusions are based on experiments in hypophysectomized rats, where the role of various hormones in starting or promoting the stages of the arrested spermatogenesis can be analysed. Spermatozoa become mobile in the epididymis after their morphological differentiation is completed in the testis, but their functional differentiation is as yet incomplete. In man, they spend 9 to 14 days in the epididymis and acquire some of their fertilizing competence there. Their final stage of functional differentiation is accomplished by capacitation in the female genital tract, in about 1 hour in the mouse and 7 hours in man. The sperm retains its fertility for about 12 hours in the mouse and 30 hours in the rabbit. Capacitation is reversed in the seminal plasma and restored in the female genital tract. While the ovum is induced to complete the second meiotic division by the entry of a sperm, the sperm acquires its full capacity for fertilization only in the female genital tract.

The sexual urge and fertility decrease in both sexes with advancing age. Man and most male mammals can produce fertile sperm even in old age. In females, the oestrous and menstrual cycles become more irregular in old age, but persist even in non-human primates. They are recognized by the reddening and swelling of the sexual skin of the genital regions in monkeys. In gorillas and orang-utans of 30 or more years, the average life span, follicles in various stages of development indicating sexual activity are found post-mortem. Only women experience the menopause, a cessation of cycles and loss of fertility, long before they reach the end of their life span. During the last century the duration of the reproductive period in women has been extended by about 3 years at the start and by about 4 years at the end, from an average age of 47 years to 51 years, which is still some 20 years short of the average duration of life. Whether the unique phenomenon of the menopause is due to an extension of the expectation of life without a corresponding conservation of ova and follicles, must remain a philosophical speculation. The menopause is supposed to be due to a reduced responsiveness of the follicular cells to the presence of gonadotropins, since ova are often still present at this time and the output of FSH and LH is increased by the lack of the dampening feed-back effect of the zero or insufficient levels of oestrogens and progesterones. The oestrogen withdrawal (or the overproduction of FSH) gives rise to a series of well-known somatic and psychological symptoms such as flushes, irritability and depression, which can be relieved by the administration of suitable doses of oestrogens. In men, some psychological symptoms may appear at around middle age. Whether these are related to a decline in sexual urges because of primarily psychological or primarily hormonal factors, has not been satisfactorily established.

In males, sexual urges decline in step with other bodily functions such as muscular activity, lengthening of the reaction times, abilities to learn and sensory perceptions of sight and hearing. This is a gradual deterioration of functions, while in the menopause the cessation of fertility is dissociated from the gradual decline with age in somatic and mental faculties. There is no evidence for an effect of paternal age in children fathered by old men, but there are abnormalities associated with maternal age, such as trisomy 21 in the example of Down's disease. There is a difference in age of mature gonocytes with sex; in males of all ages mature spermatozoa are always about 74 days old, while the ova are as old as the women, since they have been formed before birth.

The process of sex determination is initiated by the action of the sex chromosomes, but is not unalterably determined by them. Gonadal hormones and the neuroendocrine secretions play decisive roles in the differentiation of the reproductive organs and in the determination of the male and female characteristics of a species. The successive steps of the development are influenced by a variety of environmental factors which are most obvious in the onset of puberty, that of the breeding seasons, the regulation of the cyclical reproductive activities. These influences may vary with species, as for example in the influence of light, temperature, availability of food, presence of members of the same or the other sex.

The importance of differences in the maturation of gonocytes for the persistence of heritable characters, will be discussed in a later section. The chromosomes are undoubtedly the initiators of the process of sexual development, but they contain the programme merely in outline and complex interactions of the gonads, neuroendocrine centres with other endocrines and sensory organs are responsible for the execution of the programme, and are capable of altering it in detail. The biochemical and molecular properties of the sex chromosomes may affect directly the differentiation of specific cells and through them may influence indirectly the development of distant tissues and organs. The timing of the various integrated functions to produce the normal sexual differentiation may vary with species, and they are obviously very complex and incompletely understood. The mere presence of a chromosome is not sufficient to explain the sequence of events which, in their timing and in detail, are subject to both endogenous and exogenous conditions. Sex determination is thus not a single continuous process, but a step-like progression in which chromosomal, gonadal and neuroendocrine factors play the predominant successive roles.

7. Sex ratio and life span

The sex ratio is the proportion of males to females. It changes from that at conception—the primary sex ratio—to that at birth—the secondary sex ratio—to the continuously altering one in later life—the tertiary sex ratio. It is best documented in man by the population statistics collected and published by many countries. For other mammals, the most reliable figure is that for the secondary sex ratio, since subsequently the proportion is changed artificially by human or specific environmental interference, usually in favour of the female sex. This is very obvious in farm animals, where the males are slaughtered or their physiology is altered by castration. For breeding purposes, fewer males than females are required, and even in natural surroundings surplus males are excluded from the group and thus from the best feeding grounds. This is well illustrated in the social structure of a herd of deer, where a single male rules a harem and drives off weaker rivals, which may find it difficult to fend for themselves in isolation. The life of the human individual is cherished, and except in wars, social upheavals or crimes, losses are not deliberate, but due to accidents and illnesses. Human sacrifices to placate gods were frequently practised at least in populations far removed from contact with civilization. In many cultures, a boy is valued above a girl for continuing the male line of descent, as a potential hunter, warrior, protector, worker and provider of food. The attitude to domestic animals is different; the cow, ewe and sow are preferred to the bull, ram or boar for the purposes of breeding, for easier handling, the production of milk and dairy products. Hence the number of males is reduced or their physiology and psychology changed by castration.

The sex ratio of many mammals at birth is greater than 1·0, indicating that more males than females are born. Fig. 4 gives some examples, and indicates the mean value and also the range of variations. The deviations from the mean are smallest in man, where the best and most extensive data are available, and largest in the rabbit and ass. The range of fluctuations between the extreme samples tends to a higher rather than a lower sex ratio, i.e., the male preponderance at one end is greater than the female one at the other. Man is a

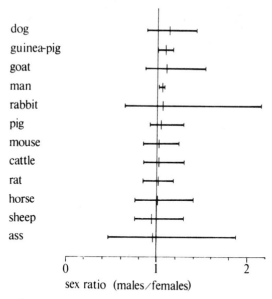

Fig.4. The secondary sex ratio, i.e. the proportion of males to females at birth in some mammalian species.

possible exception to this rule, but the lowest value is still 1·02 and thus above equality. In guinea-pigs, 1·0 is the limit at the lower end, but in other species, it falls well below equality and as far as 0·46 in the ass. If X- and Y-carrying spermatozoa have an equal chance of fertilizing an egg, the sex ratio at conception should be 1·0. The inequality of the secondary sex ratio might result from a differential loss of female embryos or foetuses during the period of gestation, or from preferential implantation of XY conceptuses. Alternatively, X- and Y-carrying spermatozoa may have an unequal chance of fertilization, either because of inherent qualitative differences or because the proportion produced in the testis varies, or because the loss of one type or the other in the male and female genital tract differs.

At present this problem cannot be resolved satisfactorily, since the primary sex ratio cannot be determined directly because of technical difficulties. Estimates are based on chromosome analysis at the first cleavage division, in blastocysts prior to implantation, or on the presence of Barr bodies in early embryonic stages. In all these investigations, only small numbers of specimens can be examined and of these some 40 to 80% are rejected as unsuitable because of technical defects in preparation. This applies to material in mice, hamsters, rabbits and pigs. Though an excess of males in hamsters and of females in rabbits has been reported, it is very doubtful whether these results, based on small numbers, can be extrapolated as typical for the species. There is no acceptable evidence either for or against equality of the sex ratio in preimplantation stages. An attempt was made some 50 years ago to relate the number of corpora lutea, present in the last week of pregnancy in mice, with the number and sex of animals of the litters born subsequently. Of 1200 litters examined, only 10% showed the same number and an equal distribution of the males and females in the litter. Prenatal losses amounted to as much as 70% and with increasing losses the sex ratio decreased, suggesting a preferential deficit of male conceptuses. Unfortunately, the sex of the aborted foetuses was not recorded, since the techniques suitable for chromosome sexing were not available at the time. In the majority of litters of mice the secondary sex ratio is greater than 1 (cf. fig. 4) and a preferential loss of males would suggest a still higher sex ratio at conception. Since prenatal losses may be due to different causes, it is impossible to assume that all of them affect males more than females and so to draw conclusions about the typical primary sex ratio on the basis of this study.

The other approach to the problem is by the sexing of aborted foetuses. About 15% of pregnancies in women end in spontaneous abortions, and of these 40% are attributed to foetal, and the rest to maternal, deficiencies. In pigs, about a third of the abortions during the first half of the gestation period have chromosomal abnormalities. The sexing of human abortuses is based on the appearance of the external genitalia. These are not very clearly differentiated during the first three months, and the data are thus unreliable. Sex ratios of 2·28 to 4·31 for foetuses aborted during this time have been quoted. More reliable figures are available for later periods, and sex ratios between 1·19 and 3·29 have been found. More recently, cytological techniques have been applied particularly to induced abortions, but the results obtained are conflicting; some authors claim an excess of males and others of females. Here again the samples are small and the technical difficulties great. New data may prove more reliable, if the methods are improved and the number in the samples increased. The best estimates accepted from the sexing of large numbers of later miscarriages and the secondary sex ratios are 1·5 for both man and pig.

An alternative hypothesis assumes an equal primary sex ratio, but preferential implantation of XY eggs related to HY antigens on the sperm. These are considered responsible for the development of a larger placenta, and thus for the more rapid growth of male than of female foetuses. The seminal plasma is antigenic and may coat the sperma-

but it is not established that this affects differentially X- and Y-carrying sperm. On the same theoretical considerations the influence of HY antisera on the sex ratio was investigated in pure strains of mice. The antiserum against the Y-carrying spermatozoa is made by females who receive repeated skin grafts from males of the same strain. The serum is used to treat spermatozoa prior to the insemination of untreated females, in the hope that the litters should consist only, or mainly, of females. This treatment reduced the usual secondary sex ratio of 1·10 to 0·83, and thus shows some degree of effectiveness, and supports the claim that immunological factors may have an influence on the sex ratio. That not all males are eliminated by the treatment is attributed to the weakness of the antiserum. In another attempt to investigate this anti-Y action the number of sperms lysed by treatment with the antiserum was assessed, and found to be in excess of the 50%, which might be expected from equal proportions of X- and Y-bearing spermatozoa. The excess was attributed to the presence of some Y-like antigens on the X-bearing sperm. Of these two lines of study, one suggests an insufficient and the other an excess loss of one type of sperm by treatment with the specific antiserum. This conflict might be resolved by assuming that, at least in the experiment of lysing, the proportion of the two types of sperm may vary at times, which would also account for the low secondary sex ratio in the other experiment, which is just at the lower limit of the range of variation in untreated litters (cf. fig. 4). Such variation in proportion of spermatozoa may account for the differences in successive litters produced by the same parents.

Because of the economic importance for the breeding of farm animals, efforts have been and are being made to detect differences between the two types of sperm, with a view to their separation and use for artificial insemination. The X-chromosome has a greater volume than the Y-chromosome, and this is reflected in a 2% difference in the DNA content and dry weight of sperm. The dimensions of the sperm heads are the same, and claims of significant variations in electrical charges, in density sufficient for separation in a centrifuge, and in other major properties have not been substantiated. Motility of the heavier sperm carrying an X-chromosome has been postulated to be slower, and tests in superimposed layers of albumin solutions of increasing concentrations have suggested that the majority of the faster sperm are of the Y type. The identification rests on the fluorescent staining with quinacrine of the Y body, a method used for the identification of the Y-chromosome at metaphase in human cells. At present only 5% of spermatogonial metaphases give a positive reaction, and only 50% of spermatids are adequately stained. Improvements in the techniques for identifying the types of spermatozoa may show whether there is a preferential loss of one or the other type of sperm at the spermatid or other precursor stages of spermatogenesis, and whether differences in their proportions vary with time and account for the variability in the sex ratio.

As an alternative possibiliy, there is some suggestive evidence for selection of sperms in mice and rabbits. If sperm carrying a genetic marker are mixed with an equal number of sperm without the marker, and the mixture is used for insemination, the proportion of young showing the genetic trait deviates from the expected 50%. The semen of some bulls used for insemination of cows has resulted consistently in a sex ratio greater than the usual 1·02 average. Maternal factors may play a role, as suggested in mice with alleles at the T-locus: mating at late oestrus raises the proportion of young with the T allele to 40% from that of 23% obtained by mating at earlier stages. Whether an increase in the sex ratio at birth from an average of 0·80 to 1·02 in strain A mice following improvements in the diet are due to maternal factors, or to the proportions of sperm produced by the male, or the preferential survival of XY conceptuses, has not been decided. A similar increase in the sex ratio has been reported for a substrain of mice differing from the original line in having a more alkaline pH in the blood. A high male preponderance in babies has been noted in some first-cousin marriages, in the Samaritan communities, and in women with blood group AB.

Cell deaths during spermatogenesis account for the deficit of spermatozoa derived from a single spermatogonium in many mammals; this loss is calculated to be 22% in the rat, 19% in the bull and 13% in the mouse. Loss of spermatogonia would not affect the proportion of the two types of sperm, but degeneration during the meiotic divisions and of spermatids could do so. These have been well documented, but there is as yet no information on whether there is a preferential loss of one type or the other. This could not be a constant feature anyhow, since successive litters born to the same parents often have different sex ratios. It would thus be a chance effect, though on average the deficit would have to be greater for X- than for Y-bearing sperm, since the primary and secondary sex ratios tend to be greater than 1. Improvements in the identification of the two types of spermatozoa and their precursors might help to solve this problem.

The changes in the tertiary sex ratio in postnatal life are best documented in man, and are poorly known even in the usual laboratory animals—mice, rats, guinea-pigs and rabbits. For domesticated and farm animals, the elimination of males by slaughter and the changes in their physiology brought about by castration restrict the data about the natural life span to females. Even these are not often allowed to survive to the limit, and this applies also to laboratory animals. Thus large-scale population statistics are available only for man and to a smaller extent for mice and rats.

Changes in the proportion of males to females in successive age groups are illustrated in the graph of fig. 5, which is based on data for the population of Denmark in the 1960s, but omits some kinks in the curve. The high sex ratio at birth declines steadily, reaches equality at around 25 years, and drops more steeply from the age of 55 years

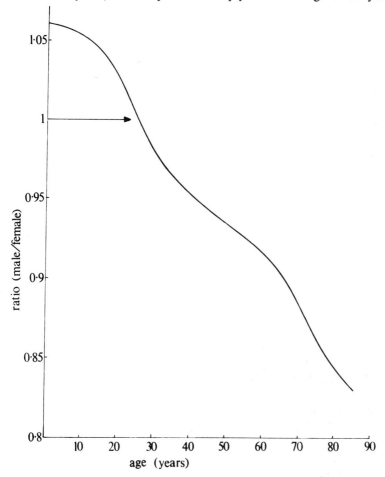

Fig.5. Changes in the sex ratio of men to women from birth to old age.

onward. Such a curve is typical for civilized countries in recent times, though variations occur from year to year and in different populations. In all of them, the essential fall in the proportion of males to females with advancing age is observed. This implies a greater loss of males than of females at all ages; it is due to a variety of causes, some of which are physiological, others due to psychological differences and social conditions. Aggressiveness is a feature of most male mammals, and is responsible for combats in sport as well as war and crime, and for taking risks in various activities. The organization of society at the present time selects men for manual jobs with high risks from accidents or from occupational diseases. We shall deal here only with the physiological characters that predispose males to contract certain diseases, some of which may be fatal.

Changes in society such as improved living conditions and nutrition, and progress in hygiene and medicine, have affected the age and sex structure of populations, particularly during the last century and a half. The average life span has increased, and the drop in infant mortality has been one of the major factors. Social conditions and the introduction of antibiotics have reduced the toll of infectious diseases and in particular of tuberculosis. The average life span of women was lower than that of men at the beginning of the period, since so many of them died in childbirth. These are some of the reasons for the extension of the average life span, and particularly for that of women. This in turn has increased the prevalence of diseases associated with advancing age, such as many cancers, heart diseases, and rheumatic illnesses, and of accidents at home and on the road through failing sight or hearing, through muscular disability and slow reaction times. All these various changes contribute to the shape of the curve for the sex ratio at various ages, but can be modified by the conditions prevailing in some countries and by the effects of an epidemic, of wars, earthquakes, floods and other disasters. The latter will make themselves felt quickly as bumps in the curve, and except for wars and some epidemics, should not affect the sex ratio. Droughts and the consequent food shortage may also be reflected quickly in the curve and alter the sex ratio, if wives and mothers forgo food in favour of men and children. Obviously the curve for the sex ratio in primitive or developing societies differ significantly from the prototype for that of civilized populations depicted in fig. 5. Rampant malaria and other infectious diseases, malnutrition, parasitic infestations and infant mortality distort the shape of the curve. The impact of the acute diseases and disasters in such societies is more immediate than that of chronic diseases associated with advancing age in the more advanced countries. The prevalence of infectious diseases in the last century and the first half of the present was reflected rapidly in the population statistics and revealed significant trends after a few years. The decrease in the death rate from acute diseases and the increase in that from chronic ones make it essential to collect such statistics over prolonged periods for a meaningful analysis of changes and their causes. The attempt to interpret data for such diseases after short periods, and even after a decade, has led to meaningless conclusions, particularly when applied to the prevalence of certain types of cancer. A close statistical correlation between the rainfall in Brazil with the membership of Welsh choirs over periods of years does not imply a cause and effect relationship. The fibre content in the food of African populations and that in Western countries is assumed by serious investigators, and more so by popularizers, to account for the difference in the prevalence of certain cancers and their increase in recent years, neglecting the age and sex structures of the populations, their different habits and general conditions of health and hygiene. The interpretations of such statistical comparisons demands supporting evidence to substantiate the assumption of causal relationships.

The graph of fig. 5 reflects the outcome of all the contributing factors for the preferential loss of males, and losses through occupational hazards such as accidents in miners and pneumoconiosis (coal dust disease of the lung) are obvious environmental factors involved. Our concern is with the predisposition to certain diseases varying with sex, where the exogenous component is less important than the chromosomal, hormonal and other physiological factors involved in sexual dimorphism. This is particularly true and demonstrable in congenital abnormalities, which are present at birth or become manifest later when they have developed and reached a state at which they can be diagnosed. The only environmental factors are those restricted to intra-uterine life, and may be attributed to endocrine or nutritional deficiencies in the mother, to maternal diseases or to the intake of drugs or exposure to damaging agents such as ionising radiations. The example of the thalidomide tragedy, and of deafness induced by German measles, are well known illustrations of untoward causes of abnormalities in babies induced before birth.

These environmental agents tend to affect equal proportions of male and female foetuses. Many congenital abnormalities are spontaneous, having no known external cause, and can often be ascribed to genetic components. Some of them are due solely to faulty single genes or groups of genes which are responsible for enzyme deficiencies. A well-known example is the deficiency of clotting factors in bleeders, causing haemophilia, where the genes are located on one of the X-chromosomes in at least one form of the disease. Such genetic abnormalities are called inborn errors of metabolism, but are not clearly distinguished from other congenital entities caused entirely or partly by chromosomal factors. If the latter are associated with the X-chromosome, they are termed sex-linked; if carried on autosomes but manifested preferentially in one sex, as sex-limited. These usually require additional hormonal or physiological conditions for their expression. The genetic component is revealed by the family pedigree showing inheritance by males from their mothers in sex-linked conditions, and in sex-limited defects by an increased incidence in siblings. When the deficiencies affect early stages in the foetal morphogenesis, such as the formation or closure of the neural tube, they are described as malformations, as for instance those leading to anencephaly (absence or malformation of the brain) or spina bifida (failure of the neural tube to close). In later foetal periods, structures already formed are involved, giving rise to deformations, such as club foot or dislocation of the femur in the hip joint.

In animals, malformations and deformations can be induced experimentally by various teratogens (drugs inducing foetal aberrations). Thus cleft lips and palates can be induced in mice by the injection of cortisone at certain stages of pregnancy, or by the exposure of the foetuses *in utero* to X-rays. By these means, all the defects recognized in mutant strains can be imitated; but while the latter are mostly confined to a single anomaly and its consequences on other organ systems, those produced experimentally have a much wider range, including the particular anomaly studied. Furthermore, while many of the spontaneous defects are sexually dimorphic, the induced ones do not differentiate between the sexes. To give one example; X-rays induce the reduction or absence of eyes in mice equally in males ad females, but in a mutant strain this condition is found in 20% of females and in only 3% of males.

Table 1 illustrates the variation in sex ratios of some common human congenital defects. Some of these are not compatible with survival (anencephaly), others are so severe that they handicap the individual for life (spina bifida with paralysis of the legs and lower part of the body), others are suitable for complete or partial correction (dislocation of the hip, club foot, cleft lips and palates). Pyloric stenosis causes muscular spasms at the entry of milk into the stomach, and thus interferes with the feeding of the baby. This can be remedied surgically, and causes no ill effects or problems later on. Down's syndrome (trisomy 21) causes a variety of symptoms including mental

Table 1 *The sex ratios of some of the more common human congenital abnormalities†*

Defect	Sex ratios (males/females)
Dislocation of the hipjoint	0.14
Anencephaly	0.3
Spina bifida	0.6
Congenital heart defects	1.0
Down's syndrome	1.0
Cleft lip and palate	1.8
Club foot	2.0
Pyloric stenosis	5.0

†The variations in the prevalence rates of the congenital disorders are independent of the sex ratios

retardation and heart defects, and also predisposes the individual to a blood cancer (leukaemia), the incidence of which is increased to 11 times that in the normal persons of the age groups. Though the sex ratio at birth is given as 1·0, this syndrome is responsible for a prenatal mortality of 65%, and of these losses 70% are of females. In the first decade of life the sex ratio for deaths drops to 0·3. This condition is thus more severe in girls than in boys both before and after birth. The same is true for hydrocephalus (water on the brain—an increased amount of fluid in the vesicles, compressing the hemispheres of the brain). This condition is twice as frequent in girls as in boys, and is manifest at an earlier age and arrested in a smaller percentage of the females.

The prevalence of defects of the neural tube (anencephaly and spina bifida) varies with district in a country, and with countries, as well as in different years, producing what are often described as epidemics. In spite of these variations, the sex ratio remains fairly constant. A study of abortuses in Japan has concluded that about 80% of the affected foetuses have died before birth. This prenatal loss may account for the low sex ratio, assuming that most of the more severely affected males have aborted.

These few examples are given to illustrate the variations in the sex ratio of some of the more common congenital abnormalities. The presence of a genetic, or at least familial, component in them is indicated by the greater than normal incidence of anencephaly or spina bifida in siblings or relatives of an affected individual. Though heart defects appear in the table as having an equal sex ratio, the localization of the fault may differ: in boys they are predominantly transpositions of the great vessels and a narrowing of the aorta, while in girls the foetal ductus arteriosus remains open.

The percentage of chromosome abnormalities in boys, at 0·3%, is twice that of girls, 0·15%, and accounts partly for the greater pre- and postnatal losses. Not all deficiencies are fatal in the short run, but they may contribute to shortening the life span. Many of the best-documented inherited defects are carried on the X-chromosome, and thus affect boys more than girls, who can compensate by the action of a normal X-chromosome in about 50% of the relevant cells and tissues. Colour blindness for red-green, haemophilia and the severe immune deficiency are sex-linked conditions of graded severity for the afflicted individual. Colour blindness is a minor handicap, and its prevalence has led to the change in the colour code for electric wiring. The position of the green and red in traffic lights allows the man with the handicap to overcome it. It is not a factor likely to affect his chance of survival. Haemophilia is a more severe condition causing pain and crippling from haemorrhages into the joints, and fatal losses of blood from injuries which would not be lethal in the ordinary person. It restricts activities so as to avoid minor injuries, and often requires lengthy treatments in hospitals. The X-linked immunodeficiency renders boys unable to cope with even minor infections; it is manifest at the age of three to six months and is usually fatal within two years. Chromosome and genetic defects are only some of the factors contributing to the increasing deficit of males in postnatal life.

The sex hormones are responsible not only for the sexual development, but also for many other physiological and metabolic processes which affect the duration of life. Female rats are smaller than the males, but live on average 10% longer. Castration prolongs the life of males, spaying shortens that of females. Testosterone given to castrates shortens their life, and oestrogens prolong the life of spayed females. These effects are related to the gain in weight and thus to metabolic factors, a point brought out clearly in observations on a pure strain of mice. If the young after weaning are given as much food as they like, when three to six weeks old two groups can be easily distinguished by the differences in the gain of weight. The group growing fastest continues to do so for the rest of their life, and to deposit more fat than the other group, but the average life span is cut short, by half in males and by a third in females. There is no significant difference in the average survival of males and females of the slowly growing mice, but in the greedy group the average life span of females is 60% greater than that of males. In hamsters, the females are larger than the males, are more active, and have a higher body temperature, but have a shorter expectation of life. The effect of sex-steroid hormones on the expectation of life in man has been analyzed in an institution for the mentally retarded in the United States. The average age at death of the eunuchs among the inmates was 69 years compared with 56 years in the intact patients. In species with male dominance the growth rate and aggressiveness are related to the action of androgens, while female dominance over other females in hamsters is reflected in the level of progesterone. Physical activity, higher body temperature and metabolic rate are considered to lead to a more rapid burning-up, and thus to a shortening of the life span.

The role of sex hormones in human diseases is revealed in changes of the sex ratio at puberty and at the menopause. Heart diseases resulting in heart failure affect more men than women. The sex ratio up to the middle forties is 13, but falls to only 2 after the menopause. In premenopausal women the oestrogens keep down the lipoprotein and cholesterol levels in the blood, and thus protect them against cardiovascular diseases, which increase with the decline in ovarian functions. The metabolism of purines, and with it the level of uric acid in the blood, changes at puberty in boys but not in girls, and may lead in the males of predisposed families to the deposition of uric acid crystals in the joints and other tissues, thus causing gout, which has a sex ratio of 19. The high incidence in some families suggests either a genetic component, or a particular style of living as regards food, drink, tobacco or occupation, or both. The genetic component may imply a predisposition to react in a certain way to exogenous stimuli, and may be polygenic, i.e. due to many genes rather than to a single faulty gene. These need not be carried on the X-chromosome, and the sex differences in the diseases of predisposed families may be due to the combination of physiological and hormonal factors with the genetic defects. The separation of the contributing causes according to their importance is often very difficult, if not impossible. In children under five years of age, whooping cough is prevalent and more severe in girls, while meningitis affects more boys and affects them more severely. In the second decade of life, poliomyelitis (infantile paralysis) is commoner in boys, and rheumatic chorea (a rheumatic infection of the brain) in girls. Thus a sensitivity to the infectious agents, or alternatively a resistance to them of a rather specific type, is somehow linked to a male or female constitution. The age of children under five years makes it unlikely that the sex hormones are responsible for the differences in reaction, and there is no obvious familial trait suggesting a genetic component. In adults, the response of organs to insults of various forms is known to differ with sex, but nervous tissue is involved in the poliomyelitis and in rheumatic chorea. In adult men, conditions affecting the heart and arteries, the lung and the kidneys exceed in number and severity that in women, while the latter tend to have more diseases of the thyroid, gall bladder and pancreas. Immunological deficiencies are more common in men, while auto-immune disorders with an overproduction of antibodies against the individual's own organs, as for instance the thyroid, are predominant in women.

The pattern of organs affected by cancers varies considerably between the sexes and with age. The primary and secondary sex organs, such as the ovary, uterus, vagina and breast, are the sites of the majority of the cancers in women, while in men, the lung, stomach and prostate are the most frequent sites. The sex ratio for cancer mortality is high in the second and eighth decade and low in the fourth and fifth, when breast cancers and those of the female genital tract reach their peak. The incidence of cancers at various sites varies with country, and the sex ratio may or may not vary with it, as the following examples show. Stomach cancer is more frequent in Japanese than in white Americans, but the sex ratios are 1·93 and 1·94 respectively; thus the increase is the same for both sexes. The mortality from lung cancer in Finland is 4·4 times that in Japan, but the sex ratio differs by a factor of 5·4, i.e. the higher incidence is due to more deaths in Finnish men. The same relation is shown in a comparison of mortality rates in Finland and Norway for 2 periods 12 years apart. The rate for Norway remains lower than that for Finland, but has increased in men by a factor of 2·90, though not at all in women. The Finnish rate rose in men by 2·75 and in women by 1·5 times. Thus the mortality has risen in males in both countries by an almost equal amount, but has risen only in Finnish women. Such figures are interpreted as demonstrating environmental causes for cancer incidence—in this instance mainly cigarette smoking—but the differential behaviour in the sex ratio may implicate endogenous conditions as well. For the action of exogenous factors it is assumed that the risk of cancers increases with the dosage of the agent, but it has not been shown that men and women react equally to the same dose of a carcinogen. Experiments in rats have shown that the same dose of a chemical carcinogen injected into the salivary gland of males and females induces significantly more tumours in males than in females. Additional administration of androgens to the females raises the level of cancer incidence to that of males, while oestrogens given to males lower it to the female level. So it cannot be assumed that males and females react alike to the same sort of treatment, and later on the same will be shown for responses to various drugs.

Other factors such as race and social circumstances condition the pattern of cancers; American negroes have five times as many breast cancers, but only twice as many uterine cancers as Japanese women, while negroes in America have ten times as many prostate cancers as Japanese. Thus genetic, racial, hormonal, environmental and the ill-defined constitutional factors are involved in the development of cancers as well as of other diseases, and their role may differ from the genetic or chromosomal predominance in sex-linked disabilities to that of infectious agents (bacteria or viruses) in epidemics. The role of social conditions amongst the environmental factors should not be underestimated.

8. Sex differences in the rate of growth and maturation

Four periods of growth and maturation in the development of mammals can be distinguished: the foetal period ending at birth, the postnatal period of childhood, puberty or adolescence, and the adult stage when growth in length ceases and maturation is complete. The sex differences in these processes vary in the four phases, and before dealing with them and with the factors influencing them, it is appropriate to clarify the relations of increase in size of the body and its constituents to their differentiation. Proliferation of cells is considered incompatible with their differentiation according to the dictum 'a dividing cell does not differentiate and a differentiated cell does not divide', which applies generally. This statement cannot be extended to the conditions in organs or the whole body, since these structures grow by the multiplication of cells, by the enlargement of differentiating cells and the deposition by them of fibres, cartilage, bone matrix and other substances. At these levels, the proliferative and functional activities of both the parenchymal and supporting cells have to be coordinated, and to change from the initial predominance of cellular multiplication to the later stages in which cell division ceases or is restricted to the replacement of cells. Cells are lost constantly in the processes of differentiation and ageing; keratinized cells are shed from the surface of the epidermis, intestinal cells from the tips of villi, erythrocytes are broken down in the spleen, and ganglion cells of the cerebral cortex degenerate. The healing of wounds, the regeneration of the liver after partial hepatectomy, or the hyperplasia of a kidney after unilateral nephrectomy require cellular proliferation. These considerations deal with the parenchymal cells, i.e., those characteristic for the tissue or organ. The supporting connective tissue is able to proliferate throughout life.

In the diagram of fig. 6, the different potentialities open to parenchymal cells are illustrated, without indicating the time relations of the various activities. In an adult tissue capable of cell renewal, such as the epidermis or the intestinal epithelium, the basic or stock cells typical of the tissue are designated as Go (cf fig. 1). They may be functionally relatively inactive, or like glandular cells they may be

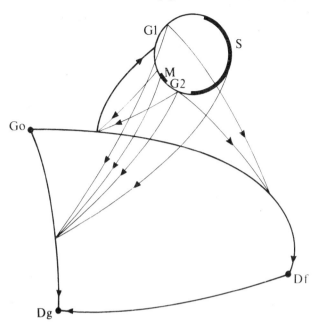

Fig.6. The parenchymal cell cycle in mammals.
The basic or stock cell of a tissue (e.g. basal epidermal cell, crypt cell in the intestine, liver cell) is a potentially dividing cell (Go) and may function as an *intermitotic* differentiated cell in glands and the liver. It may differentiate and lose its capacity for mitosis (*postmitotic* differentiated cell, Df) becoming, for example, a ganglion or a keratinized cell, or may degenerate (Dg). Cells in Go and the cycle (G1, S, G2, M) form the proliferating compartment of the tissue and may differentiate in Go, G2 or in G1 after completing division (fig.1). All cells may degenerate (faint lines) and some forms of differentiation (keratinization, haemoglobin formation in erythrocytes, sebaceous cells) end in death of the cell. In a renewal system the number of cells lost is replaced by newly formed cells and in the simplest form one of the daughter cells of a mitotic division remains in the cycle, while the other differentiates. In a growing organ the proliferative and functional compartment must increase, and this is accomplished by changing the proportion of proliferating to differentiating cells from a ratio of 1 to 1 to 1 to 3 or more. Differentiating cells have a longer life span than cells in the cycle and thus the contribution of single cells to the functional is greater than to the proliferative compartment of the tissue. In some adult organs (brain) multiplication ceases. Degenerating cells remain visible for 1 to 7 hours unless present in very large numbers.

engaged in specific secretions, but capable of reproduction, and are intermitotic differentiated cells. Potentially dividing cells become actively so when entering the cycle via G1, S, G2, M and as daughter cells in G1 may continue to cycle, to return to Go, or to leave the cycle to differentiate in either G1, G2 or Go, when they enter the functional compartment and leave the proliferative compartment made up of Go and the cycling cells. Whatever stage they are in, cells of both compartments may be lost by degeneration at the end of their life span (keratinized cells, erythrocytes, surface cells of the intestines) or may be killed by physical, chemical or biological agents. Discarded cells are replaced in a renewal system by cell proliferation, and the example of inciting mitotic activity in the basal layers of the epidermis by stripping surface cells with adhesive tape has been mentioned before. Losses of cells and formation of new ones are balanced by one of the daughter cells of a mitosis remaining in the proliferative compartment and the other starting to differentiate. There is thus a 1:1 relationship, though it should be realized that the life span of a differentiating cell may be more than 20 times the duration of the proliferative cell cycle. In adult renewal systems, the life span of differentiating cells varies from a few days in the small intestine, to weeks in the epidermis, to months for erythrocytes, to years for hairs in the human scalp, and even more years for cells of the liver. Thus the relationship between members of the proliferative and functional compartments varies in its quantitative aspects with the organ. The central nervous system and striated muscles are not such renewal systems; in them cell division ceases in the perinatal period and all cells are in the functional compartment, as postmitotic differentiated cells. Their loss cannot be compensated by parenchymal cells, and defects are filled by the supporting cells of the glia or connective tissue in the formation of scars.

Renewal systems are designed for maintaining a status quo and not for growth, so that the proliferative and the functional compartments are and remain in balance. For the growth of organs both compartments have to enlarge by increasing the number of cells and, in the case of differentiating cells their size also. Since the latter have a longer life span, the contribution of a single cell to this compartment is greater than to the proliferative compartment. This is compensated by altering the ratio of daughter cells entering into differentiation from the 1:1 in the adult renewal system, to at first 1:0 and later to 1:3 or more. Thus during different stages of development progressively more daughter cells will leave the cycle of multiplication and enlarge the differentiated cell number. This simple regulation of the quantitative relationships is slightly modified by the death of cells even in early embryonic development. Cells degenerate in the process of in- and exfolding for the development of the neural tubes, the lens and the eye, and in the joining of bilateral contributions to the formation of the palate and the sternum, and have been termed 'morphogenetic degenerations'. Some rudimentary structures regress by death of cells, and these may vary with sex, e.g., the Mullerian ducts in males and most of the mesonephros and its ducts in females. The death of many oogonia during the foetal period has been mentioned before.

Cell numbers in different organs and organ systems may cease to increase at any stage in the 4 periods of growth and maturation. Human oogonia reach their peak value in the 5th month of the gestation period, the number of cells in the striated muscles of sheep is fixed by the 100th day of gestation, that of ganglion cells in most mammals in the perinatal period. The weight of organs such as the brain and the muscles is augmented by the formation of fibres, and that of the ovary by the development of the interstitial tissue. These activities imply differentiation and maturation of the organs, and as these change in their composition, so does the whole body in respect of the relative sizes of the organs. Growth in length is due to the elongation of the skeleton, and is limited by the ossification of the epiphyseal cartilage plates which progresses in adolescence and is complete in early adulthood in many, though not all mammals. Subsequently, an increase in weight of the bones is due to an increase in thickness, and internal remodelling of the bone structure in response to mechanical stresses continues throughout life. Thus the composition of the body changes, and does so at different rates in males and females. Though the simplest measure of comparing such changes is by weight, it is not an entirely satisfactory guide, particularly since the deposition of fat and of connective tissue varies with sex from puberty onwards, and so does the rate of maturation of organs even in the foetal period. The sex ratio for the gonads in mice at birth is 4·22 (i.e. the testis weighs about 4 times as much as the ovary) and at 4 weeks it is 25·84, while the sex ratio for the weight of the whole body remains the same.

At birth, the weight and length of the body give the same sex ratio in most mammals and may thus be used as an approximate guide to the growth rate in the foetal period. In man, dog, rat, cattle, goats and sheep, the males are heavier at birth than the females, while the reverse holds for cats, mice, golden hamsters and rabbits. The sex ratio for weight is similar to that for length and is about 1·06 in rats, guinea-pigs, cattle and man, about 0·94 for cats and 0·99 for mice. In cattle, the sex ratio varies with breed from 1·02 for Ayrshires to 1·14 for Holsteins. The degree of maturation as judged by the development of ossification centres, reflex actions, and the coordination of movements is more advanced in girls than in boys by 2 weeks at 20 weeks of pregnancy and by 4 to 6 weeks at birth. Though larger, baby boys are less mature than baby girls, indicating that relatively more cells remain in the proliferative compartment than leave for differentiation. The stimulation of growth is due both to better nutrition and to androgenic secretions. The placenta of male foetuses is usually larger than that of females and androgens promote growth of the body as well as of the gonads. The influence of nutrition is highlighted in foals of crosses between Shire horses and Shetland ponies. In Shire mares the foal is three times the size of that in Shetland mares, as might be expected from the differences in size of the uterus and the maternal contribution to the placenta. The genetic contribution in the crosses may be assumed to be very similar and thus not to account for the size difference.

The rate of foetal maturation varies with species, as shown by the status and behaviour at birth; new-born rats and mice have their eyes still closed, while piglets, calves and lambs can see, stand up and suck; guinea-pigs are fairly mature at birth. Human babies are unable to fend for themselves, though their eyes are open. Duration of pregnancy, the weight of the baby as percentage of that of the mother, and the size of the litter govern the rate of growth and maturation. Some relevant data are given in table 2, which indicates that small mammals

Table 2 *The duration of the gestation period, the average litter size, the average maternal weight and the weight of one neonate as percentage of maternal weight*†

Species	Duration of pregnancy (days)	Litter size	Maternal weight (kg)	% Weight neonate/maternal
Mouse	20	7·4	0.03	3·5
Rat	21	9·0	0·24	2·2
Rabbit	31	8·0	4	1·6
Guinea-pig	68	3·5	0·8	12·2
Sow	114	9·0	88	1·4
Ewe	145	2·0	54	7·6
Woman	281	1·0	55	6·2
Cow	284	1·0	480	5·2

† The data refer to piebald mice, Wistar rats, New Zealand White rabbits, random bred guinea-pigs, Yorkshire sows, Hampshire ewes, English women and Jersey cows.

have a shorter gestation period than the larger species, a bigger litter and a lower ratio of weight of a single baby to that of the mother. There are, however, some major exceptions; the guinea-pig is smaller than the rabbit, but has a gestation period twice as long and a neonate/maternal weight ratio nine times as great. The sow has a shorter gestation period than the sheep, which is of similar size, and a considerably smaller neonate/maternal weight ratio. Human babies and calves have about the same gestation period and a comparable neonate/maternal weight ratio, but differ considerably in their maturity at birth. Obviously, inherent factors are involved and they are interpreted as being genetic—an explanation which though true is not illuminating, since by definition species differ in their genetic make-up. Compared with the interspecific diversity the intraspecific sexual dimorphism appears almost negligible.

The postnatal rate of growth varies with species and sex, and may continue or reverse the prenatal sex difference. Again smaller species tend to grow faster than the larger ones. This is shown in table 3,

Table 3 Species and sex differences in postnatal growth

Species	Sex	Weight at one week/weight at birth
Cat	male	1·32
	female†	1·39
Dog (Beagle)	male†	1·77
	female	1·74
Hamster	male and female	2·12
Rabbit	male and female	1·25
Rat	male†	2·88
	female	2·82
Mouse	male	3·23
	female†	3·20
		Weight at three months/weight at birth
Human	male†	1·68
	female	1·65
Cattle (Holstein)	male†	2·18
	female	2·44
Goat (Saanen)	male†	4·17
	female	4·65
Guinea-pig	male and female	4·94
Sheep	male†	5·09
	female	5·42
Pig	male and female	9·60

† denotes the heavier sex at birth

where the gain in weight for the first postnatal week is listed for the smaller species and that in the first 3 months for the larger animals. The time difference (a factor of 13) is of importance. The exceptions of guinea-pig and pig seen in table 2 reappear in the present data. The rabbit more than doubles its birth weight in one week, while the smaller guinea-pig puts on 5 times its birth weight in 13 weeks, so that there is a difference between the species of times 6. The pig grows twice as fast as the sheep in 3 months, doubling its birth weight in about 1 week. The larger and more mature animals at birth have thus a lower postnatal growth rate. The sex differences in weight at birth persist in the postnatal period in cats, dogs, rats and man, but are reversed in mice, cattle, goats and lambs. The data given for cattle, goats and lambs are for single offspring, and in sheep the sex differences between the sexes disappear in larger litters. With multiple births (cats, dogs, rabbits, rats) members of the larger sex at birth may have an advantage in the competition for nipples over the smaller siblings. The reversal in growth rate in mice may be due to the greater aggressiveness of the male. Injection of androgens into female mice and rats during this period accelerates their growth. The amount of food available, i.e., the milk production by the mother, is a major factor, as is shown in the comparison of the gain in weight in small and large litters. The sex differences in the postnatal rate of growth are small and in line with the cessation of androgen production in males, while oestrogen production has not yet started.

In puberty, a spurt in growth and maturation precedes the onset of the menarche and of spermiogenesis, and usually starts in females earlier than in males. Thus for a time at least the sex ratio for the gain

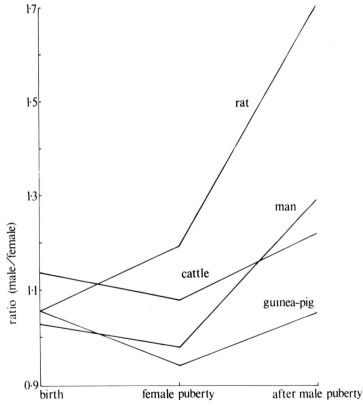

Fig.7. The ratio of male/female body weight at birth, at female puberty and at male puberty in man, rat, guinea-pig and cattle.

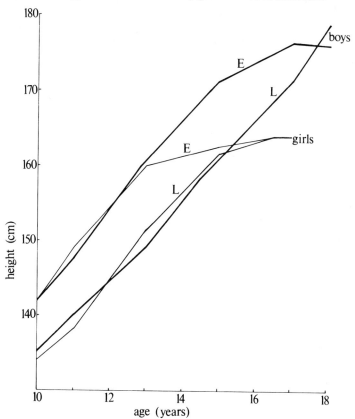

Fig.8. The height of boys and girls in relation to the onset of puberty. E denotes early onset of puberty, i.e. in boys before 14 years, in girls before 13 years, L denotes late onset of puberty, i.e. in boys after 15 years, in girls after 14 years.

in body weight changes, depending on the interval between male and female adolescence. This is illustrated in fig. 7 which shows the variations in the sex ratio for body weight in man, the rat, cattle and guinea-pigs. Puberty in girls at the age of 13 and 14 precedes that in boys by about 1 year; in the rat, the time difference is very small at around 10 weeks, in guinea-pigs, it is only 1 or 2 weeks at 2 months, and in cattle, the same interval at 47 weeks. The best data are available for man, and show a drop from above 1·0 to below equality at the female puberty and a steep rise with advancing male adolescence. The same general pattern is found in cattle and guinea-pigs, while in the rat the sex ratio stays above 1·0 and increases steeply with male puberty.

Human puberty starts earlier in taller and heavier boys and girls than in the smaller ones. This is shown in fig. 8, in which heights of two groups of boys in puberty before 14 years (E = early starters) and those after 15 years (L = late starters) are contrasted with each other and with girls in menarche before 13 years (E = early menarche) and those in menarche at 14 years or later (L = late menarche). Similar curves are obtained for weight. In the early group, boys and girls start with the same height at 10 years, then girls are taller until the boys overtake them at about 12½ years when the growth spurt in girls levels off, while that in boys remains steep for another 2 years. In the late group girls outgrow boys from about 12 to 15 years, while boys continue to grow for a longer time. In both sexes the late starters catch up with the early ones, but the growth spurt lasts longer in males than in females, levelling off at 17 to 18 years. The longer duration of the growth spurt accounts for the greater difference in height at 17 than at 10 years. From that time onwards, the height increases at a greatly reduced speed until the epiphyseal cartilage plates are ossified, which happens two years earlier in females than in males. The body stem of boys at 8 to 9 years is longer, at 11 to 14 years shorter, and subsequently again longer than that of girls. The proportion of the length of the limbs to that of the trunk is smaller in girls than in boys. Improved living conditions have led to a gain in average height of the whole population of the developed world, have decreased the differential between the sexes, and have changed the proportion of the trunk to limbs in women. Compared with previous generations, girls have become longer-legged, which means that the increase in height is due largely to that of the limbs. Other prominent physical changes at puberty are the deposition of fat on the arms and legs of girls aged between 8 and 12 years, while boys of the same age actually lose fat and develop muscles. The growth of the breast and of the genital organs characterize female adolescence, while enlargement of the testicles, of the larynx resulting in a change of the pitch of the voice, and the replacement of the vellus by terminal hairs on the face and chest are prominent features of male adolescence.

The pubertal growth spurt is associated with the increased activity of the gonadal secretions, and so with that of the pituitary, hypothalamus and pineal body, and through them with that of peripheral endocrines. Androgens accelerate growth, but also the ossification of the epiphyseal plates. Men castrated before puberty have excessively long limbs and a female configuration of the pelvis. In farm animals castration delays the ossification in the skeleton, but reduces the rate of growth of muscles, which is attributed to the reduced plasma level of androgens. The growth rate of sheep decreases in the order rams, castrates, females, spayed animals. Oestradiol promotes the growth of castrates and spayed sheep, but reduces it in intact males and females.

It speeds up ossification more than do androgens, as is seen also in the duration of the growth spurt of boys and girls. The effect of the gonadal secretions on growth varies with species; ovariectomy of rats promotes the gain in weight and length, castration of the golden hamster allows him to reach the size of the larger female. Androgens, like oestrogens, influence both the rate of growth and that of maturation, actions which are to some extent antagonistic and vary with species, with tissue and with its site and stage of development in the same individual. Androgens specifically promote the growth and development of the early maturing groups of muscles of the head, neck and shoulder in cattle, sheep and rats, those of the temporal muscles in the guinea-pig. These regions are specially affected by castration, and respond to androgenic treatments in females and castrates. These actions are exerted on the tissue itself by stimulating synthesis of RNA and protein, and deposition of glycogen and creatine in the muscles, thus favouring the development of muscle over the deposition of fat, an effect which has been described for boys and girls aged 8 to 12 years. Bulls convert a measured amount of food more efficiently into meat than do cows or castrates. Androgens act directly even in isolated tissues.

Oestrogens also promote growth, but do so indirectly by their effect on the pituitary, except in their specific effects on the tissues of the genital tract. They stimulate the enlargement of the pituitary, which is larger in females than in males; since its secretion is proportional to its size, the female pituitary produces more growth hormone than the male, resulting in higher blood levels. The level of growth hormone is one of the factors influencing growth, and since the size of the body increases faster than that of the pituitary from birth to adulthood, the concentration of hormone decreases with age. Males, having a lower level than females, respond better to the administration of growth hormone. The level fluctuates during the day, is high at night in children when the demand for fuel is low, and varies with the level of glucose. The hormone increases the uptake of amino-acids in the tissues thus causing a fall in plasma proteins; it mobilizes lipids as fuel, and so saves proteins for growth. Fasting and doses of insulin lower the glucose level, and so activate the release of growth hormone. Insulin, like growth hormone, stimulates the synthesis of proteins and so of growth; growth hormone acts by cellular proliferation, insulin by increase in cell size, but the reaction to glucose levels in the plasma differs: a low level increasing release of growth hormone, a high level, that of insulin. Excess secretion of growth hormone causes a disproportionate enlargement of the face, hands and feet (acromegaly), too little may cause dwarfism, which may be corrected by the administration of the hormone. Oestrogens promote the growth of farm animals by stimulating growth and increasing the RNA/DNA ratio of the pituitary, while androgens have no parallel effect on this organ. It is possible that the pubertal growth spurt of males and females is due to different arms of the endocrine system. Whatever it is that initiates puberty in females, it tends to do so earlier than in males; it may affect primarily the growth hormone production in the pituitary, and either at the same time or subsequently the oestrogen secretion in the ovary, which in turn stimulates the pituitary. In males, the primary and predominant effect of the initiation of puberty may be on the pituitary gonadotropins, and through them on the secretions of the testis, which are more directly responsible for the growth spurt, probably abetted by growth hormone and other endocrines. Of the latter, the thyroid affects growth, which is stunted by both hyper- and hypo-activity in the gland. It affects also the growth of hairs in some regions, but androgens are responsible for the growth of hairs in the face and on the chest, and also for the loss of hairs in the sex-limited baldness of men and male monkeys, and for the structure and function of sebaceous and salivary glands. The sex hormones in both sexes act in concert with other endocrines regulated by the hypothalamus and dependent on the cortical regions of the brain as well as on nutrition. The effects of the pubertal changes are not restricted to the reproductive organs, but are manifest also in apparently unrelated organs such as liver, adrenal, kidney, bone marrow and thymus as will be discussed in a subsequent chapter. The results of the action of sex steroids vary with species for reasons as yet poorly understood.

The fourth phase in growth and maturation is reached in the adult when the growth in length ceases. There are some exceptions, noticeably the male rat which enlarges throughout life. The gain in weight of mature mammals is due to the deposition of fat, of muscle fibres, connective tissue and other substances. In renewal organs, cells multiply to replace lost cells and not as a growth activity. This poses the problem of how the average size of a species is fixed, and the related question of the size differences between the sexes. The size of a species is related roughly to the life span, duration of pregnancy and period of growth; a horse obviously takes longer than a mouse to reach its adult size and produce a viable baby, though both species start as a fertilized ovum. The number of cell generations or, in other terms, the number of doubling of cells, in the horse must thus be very much greater than in the mouse. Cellular proliferation for growh is limited, and this limitation may be inherent in all cells of a species. If that is the case, the number of cell generations should be correlated with the life span of the species in isolated cells grown in tissue culture. On explantation and growth in a culture medium, normal diploid cells tend to die out after a number of generations, unless they undergo a malignant transformation and can be continued indefinitely. Table 4

Table 4 *Maximal life span, average weight of male, sex ratio of weight and maximal number of diploid cell generations in culture.*

Species	Maximal life span (years)	Average weight of male (kg)	Ratio of weight male/female	Maximal number of diploid cell generations
Man	110	70	1·17	60
Horse	50	445	1·06	82
Rhesus monkey	34	11	1·40	60
Cat	28	4·23	1·16	92
Rabbit	14	3·73	0·91	70
Mouse	4	0·036	0·99	28

shows the maximal number of cell duplications of normal diploid cells of different species, and compares them with the maximal life span, the average weight of the male, and the male/female weight ratio. The mouse is indeed the smallest, has the shortest life span and smallest number of diploid cell generations in culture, but while the life span of man and mouse differ by a factor of 27, that in weight is over 2000 and in number of cell generations a mere 2·0. For cat and mouse the comparable factors are: 7, 1175, 3·3 and the cat cells remain normal in culture for more generations than the human, though the life span is only a quarter and the weight and so the size only a sixteenth. The same differences emerge form a comparison of the other animals, and are supported by data from other species, in particular marsupials. Inherent differences in limiting the number of duplications of normal cells, which in the case of the data quoted are fibroblasts derived from various organs, do not determine the size of a species.

The size of mammalian cells is fairly uniform, and has no bearing on the size of the species. In fact, the cells of a mouse are larger than the homologous ones of the rat. The number of cells differs with species, and its weight is a rough guide to it. The duration of the proliferative cell cycle is also fairly constant; mitosis varies between 30 and 60 minutes, DNA synthesis by a factor of at most 4 for a time of 4 hours. The variations of G1 and G2 are larger, but not sufficient to account for the increases in turnover time in the smaller species, which would be by a factor of 2000 for mice in comparison with men. The life span of an intestinal cell from formation in the crypt to extrusion into the lumen is about the same in mice and men. The hairs in the human scalp elongate by about 0·35 mm per day, slightly more in women than in men, reaching a length of about 500 mm in a growth phase lasting 4 years. In mice, the growth phase of the hair cycle is about 18 days, and the hair reaches a length of about 7 mm, a rate of elongation not widely different from that of the human hair. The human hair follicle, where the growth takes place, is larger than that of the mouse, and the growth phase lasts considerably longer. This implies a greater number of cell generations and a greater fraction of cells remaining in the proliferative compartment (fig. 6). If more cells remain in this compartment, fewer generations are required to produce a given total number of cells. The potential for the reproduction of diploid cell generations is not the limiting factor for the size of the species, it is rather the loss early on during the growth phase of proliferating cells on their entry into differentiation, which is seen in the shorter gestation period (with some exceptions), the earlier adolescence and maturation, and the shorter life span of the smaller species.

The growth potential varies between organs and structures, with species and within a species with sex. Some of these differences have been mentioned early in this chapter, and others will be discussed in the next section. The sex differences in total size of the body (table 4) are very small compared with those between species. They point to some factors influencing the rate of growth and maturation within a similar genetic framework and nutritional and general environmental conditions. The influence of the whole endocrine system on the pubertal spurt in growth and maturation has been stressed and is verified experimentally by androgenic or oestrogenic treatment of males and females respectively in a critical period before or shortly after birth. Injection of testosterone into pregnant monkeys delays the onset of menarche and of the growth spurt of the babies, and alters the differentiation of the genital tract, the neuroendocrine centres and the behaviour of the animals. Similar results are obtained by the treatment of female rodents within the first 10 days of postnatal life, though, depending on the dose of testosterone, puberty may be precocious and the rate of growth increased. Male rodents given oestrogens during this phase have a reduced rate of growth, and impaired differentiation of the genital tract and spermatogenesis. Oestrogens given to pregnant mice inhibit the secretion of androgens in the foetal testis. The effect of androgens and oestrogens on the ossification of the epiphyseal plates has been mentioned above.

An adequate food supply is necessary for growth, and the effect of the limitation of milk in large litters of sheep on the rate of growth, the effect of inadequate nutrition on the onset of puberty in marsupials, and that of starvation in various human populations on growth and development are proof of this fact. Food supply must be adequate in quantity and quality, especially of proteins. In some regions, the lack of iodine may cause mental and physical retardation, and it is compensated for in many countries by the addition of iodide to the table salt. Whether such environmental factors and the consequent changes in the endocrine and metabolic systems have played a role in the evolution of the size of a species is a matter for speculation.

The significance of the male/female size difference within a species is a very puzzling problem. The larger sex is often, though not always, dominant, but size is of advantage only in the competition with members of the same sex for mating and procreation. Size coupled with aggressiveness, agility, cunning and perseverance contribute to the dominance of a stag over its rivals. The smaller size of the does may make them submissive, though they accept the male only when on heat and actually excite and invite them by their signals, including pheromones. The female spotted hyena is larger than the male and the most aggressive becomes the leader of the pack. The males are invited and accepted by the females only when in oestrus. Dominance and size difference are thus not confined to one sex and are factors in the organization of an animal society. The golden hamster is a solitary creature and so there is no obvious social reason for perpetuating the size differences between the larger female and the smaller male. While some of the endocrine factors involved in mediating sexual dimorphism in size within a species are understood, the significance of the phenomenon is obscure.

PART II
SEXUAL DIMORPHISM IN ADULT MAMMALS

9. Differences in the proportion of the body and its organs

Differences in the shape and appearance of males and females

In some species, males and females can be easily distinguished by their shape and appearance, while in others they are very similar. The male body is somewhat like an egg with the blunt pole at the anterior or superior end, while that of the female has the blunt pole at the posterior or inferior end: males have broader shoulders and necks, females a broader pelvic region. The female pelvis accommodates the entire genital tract and must have room for the pregnant uterus, while in the male the testicles are exteriorized permanently or at least during the breeding period.

These variations in the proportions of the trunk are less obvious than some secondary sex characteristics such as the manes of lions, the antlers of stags, the larger tusks in elephants and boars, the dorsal crest in boars, the beards in men and some monkeys. Males and females are very different in size, for instance in rats and ferrets. Pregnant and lactating females are easily recognized by the conformation of the abdomen, the breasts, and the attention to the babies. Male characteristics are located in the larger and broader head, female features in the thoracic and abdominal region. These differences become obvious only after puberty; before that period it may be difficult to determine the sex of an individual. Inspection of the external genitalia may not be conclusive because of their incomplete development, and can be difficult, especially where there is a dense hair coat in ferocious animals. The difficulties of sexing the giant pandas at the London Zoo at present may serve as example. In a minority of cases, such as the freemartins and persons with abnormal numbers of sex chromosomes (Kleinfelter's or Turner's syndrome), the appearance of the external genitalia is not a correct indicator of the true sex.

The shape of the body is modelled in the first place by four of its constituent parts: the bones, the muscles attached to the skeleton, the subcutaneous fat layer, and the skin with its appendages of hairs, glands and horns. Sex differences in these components individually and in combination account for the differences in appearance.

Variations in the structure and function of the internal organs are reflected in those of the four main modelling components, but are not readily visible. The broadening of the female pelvis affects the location of the hip joints, of the axial and appendicular skeleton, of the distribution of the attached muscles and so of the posture and gait. The hip-wiggling walk of girls and the narrow waist in the hour glass figure are partially conditioned by the width of the pelvis. The body stem of women is proportionately longer and the limbs shorter than in men, and reduces the sex ratio for the total height to 1·07 from 1·09 for the length of the limbs. The sex ratio for the shoulder region is 1·08 and tapers to 1·0 for the pelvic region. The head and neck are usually larger and thicker in males than in females, as is easily seen in bulls. The males are not larger than the females in all species; the spotted hyena, the golden hamster and the rabbit are examples where the females are larger.

The skeleton accounts for about 7% of the weight of the body and is relatively heavier in males than in females. Males and females differ in size and weight. The organs of the larger sex tend to be heavier than those of the other sex, but may form a smaller fraction of the total body weight. To determine the sex ratio for organs, their weight has to be expressed as percentage of the body weight for males to that for females. If the ratio is greater than that of the body weight the organ is relatively greater in males.

In mice, the sex ratio for the weight of the skeleton is 1·06, while that for the total body is only 1·02. The skeleton is thus relatively heavier in male than in female mice, i.e. forms a greater proportion of the total body weight. In cattle of the same live weight, bones are responsible for 7·2% of total weight in bulls, for 6·8% in steers and for 6·4% in cows. In guinea-pigs the sex ratio is 1·08 for the weight of air-dried bones of the legs, and the same for the length of the limbs. After puberty the bones of men are heavier, denser and contain more mineral ash than those of women. The osseous tissue is not inert, and is linked in metabolism with that of the kidney, regulated by the parathyroid and the thyroid amongst other organs. It acts as a depot for various salts, such as ingested radium or thorium. Sexual dimorphism in bone metabolism is evidenced by the differing prevalence of various abnormalities; osteomalacia (a depletion of the mineral content leading to softening and deformations of the bones) occurs predominantly in women and often after a pregnancy. Adolescent scoliosis (a torsion of the spine leading to the appearance of a hunchback) is an abnormality in girls at puberty, rheumatoid arthritis of middle age affects more women than men, while gout and ankylosing spondylitis (calcification of the intervertebral cartilage) are found mainly, though not exclusively, in men of predisposed families. In some strains of mice osteoarthrosis affects males, is promoted by testosterone given to growing animals of either sex and is inhibited by oestrogens administered to young adult males. In another strain of mice, osteoporosis (rarefaction of the bones) affects mainly females. The symphysis pubis of rodents and women relaxes in late pregnancy, and thus widens the port through which the babies have to pass. In dogs, rats and mice, a bone is formed in the penis and a similar structure can be induced in the clitoris of females treated as neonates with testosterone. Sex and sex hormones thus affect the formation and the metabolism of bone.

The respiratory movements of men are predominantly by contraction of the diaphragm and thus of abdominal type, while in women they are of costal type, by movements of the rib cage.

Muscular development in males of many species including man is better than in females. Bulls convert measured amounts of food more efficiently into muscles (meat) than do cows, and men tend to be stronger than women, which is reflected in their segregation for athletic performances. That anabolic drugs derived from androgens are effective in building up the muscular apparatus is shown by the ban on their use by athletes. Not all groups of muscles respond equally to testosterones produced by the individual male or administered to females or castrates. The groups responding best, and thus showing the greatest sexual dimorphism, are those of the neck and shoulders in men, rats, and bulls, the temporal muscles in guinea-pigs, and those of the legs and arms of boys and girls. Testosterone acts by stimulating the synthesis of protein for muscle fibres and the deposition of creatine. Some interesting data about the composition of muscles have been gleaned from a comparison of male cattle twins of which one has developed as a bull and the other has been castrated. The muscles of the bull contain 74% water, 21% protein, 3% fat and 1% ash, while those of the steer have 73% water, 20% protein, 6% fat and 1% ash. The forequarters of the bull are made up of 36% muscle, 10% bone and 6% fat compared with 31% muscle, 10% bone and 9% fat in the steer. The hindquarters of bull and steer do not differ in the amount of muscle (33%) and bone (8%), but do as regards fat: 5% in the bull and 7·5% in the steer. The muscles of the anterior part of the body are thus better developed in the male than in the castrate and the fat content is greater in the latter in the anterior and posterior part of the body.

A comparison of bulls, cows and steers shows that the bulls grow faster and have more bone and muscle and less fat than steers and cows. Bulls reach a live weight of 386 kg in 361 days, steers 377 kg in 383 days and cows 346 kg in 398 days. Muscles account for 38% of the weight in bulls, for 33% in steers and for 31% in cows; bones for 7·2% in bulls, 6·8% in steers and 6·4% in cows, while the fat content increases from 12·4% in bulls to 16·5% in steers and 18% in cows. The steer has ranking for growth, muscles, bones and fat intermediate between bulls and cows. This applies also to an analysis of the various muscle groups making up the total muscle weight: those around the spinal column are roughly the same at around 12·1 to 12·3%, those in the abdominal wall increase from 9·7% in bulls to 10·9% in steers and 11·5% in cows, while those in the thoracic and neck region decrease from 12·5% in bulls to 10·1% in steers and 9·0% in cows. The muscle groups develop in the calf at different times; first are those concerned with standing and walking, in the distal parts of the limbs, and with sucking in the jaws, while those of the abdominal walls are the last, since at an early stage the abdominal contents require little support.

These data are available as they are of importance for animal husbandry and the meat trade. The customer, like Jack Spratt and his wife in the nursery rhyme, may prefer a greater meat or greater fat content and this will determine whether bulls or cows or, in the middle range, steers should be produced for the market. Where cattle are allowed to graze in the fields as in England, steers are preferred for the production of meat as they are easier to handle than bulls and have more meat and less fat than cows. In Scandinavia and many continental countries, cattle are kept in stables, and in Sweden, 53% of cattle slaughtered are bulls, 14% steers and the rest females. The sex differences appear after puberty and this fact and the comparison of the bull-steer identical twins show the importance of androgens for the development of muscles.

The adipose tissue acts in a dual function as a reserve of nutrients and as a structural element in modelling the shape of the body. The rounded shape of the face, body, arms and legs of women is due to fat, and the loss of fat by emaciated persons and cachectic patients accounts for the hollow cheeks and sunken eyes. There is a sex difference in the preferred sites for fat deposition; women accumulate it in the buttocks, arms, legs and internal depots, men develop a paunch. Female rats store fat in the genital region, males in the perirenal depots. Female mice have gonadal and inguinal fat stores, males perirenal. Cattle are judged for the market by the conformation of the body, i.e. the thickness and smoothness in the appearance of the back and loin, which is due mainly to the deposition of fat and is greater in cows than in steers and least in bulls.

The rise in the fat content of the body is due to (a) an increase in the number of fat cells, which in man ceases at about 25 years, (b) an enlargement of the adipose cells by the accumulation of fat and (c) a deposition of lipids in the parenchymal cells of the liver, heart and internal organs. Fat deposition increases with age and varies with sex. The ratio for fat as percentage of total weight for men to women increases from 0·54 at 25 years to 0·66 at 55 years. The fattening process in cattle accelerates when cows weigh 60 kg, steers 75 kg, bulls 85 kg. Spaying and castration promote the accumulation of fat in probably all species. Apart from the sex hormones, genetic factors and other endocrines as well as nutrition influence the metabolic rate and with it the accumulation of fat. Some strains of mice are genetically obese, females more so than males. In some families, obesity of menopausal women is linked with diabetes. Obesity is not necessarily due to gluttony, and may be caused by hormonal and/or genetic factors as some fat persons do not eat much. A metabolic malfunction enhances the conversion of food into fat instead of proteins and carbohydrates, a pathway more common in female than male cattle, pigs, sheep, rats and mice.

The colour and firmness of fat deposits vary with species and in individuals with site. In sheep, the fat is whiter and firmer than in cattle and even more so than in pigs. To some extent the colour varies with the food. The perirenal fat deposits store the firmest and the subcutaneous ones the softest adipose tissue. The influence of sex hormones on the composition of fat in the various sites is shown in table 5 for the identical bull and steer twins. In both, the water and

Table 5 *Variation in the composition of fat in various depots of identical male twins in cattle, one of whom is a castrate*

Depot	Water (%)		Protein (%)		Fat (%)	
	Bull	Steer	Bull	Steer	Bull	Steer
Subcutaneous	24·24	17·23	9·19	6·10	66·65	76·69
Intermuscular	24·93	22·71	7·20	6·57	68·16	70·94
Perirenal	7·08	5·48	1·46	1·11	91·66	93·39
Mesenteric	17·24	14·71	3·38	2·75	79·43	82·68

Based on data quoted by R.T. Berg and R.M. Butterfield (1976) *New Concepts of Cattle Growth,* Sydney, Sydney University Press.

protein contents are lowest and the fat highest in the perirenal depot, and the mesenteric fat tends to have a similar, though not such extreme, composition. The intermuscular and subcutaneous deposits have more water and protein and less fat than the internal sites. The bull and steer differ, the former having more water and protein and less fat at all sites. There is some correlation between the consistency of the adipose tissue and the temperature gradient in the body; the percentage of lipids rises as the temperature gradient between the external and internal regions increases.

The deposition and utilization of fat are controlled by various agents in addition to the sex steroids. They may promote the mobilization of fat (lipolysis) by acting on a hormone-sensitive lipase via cyclic AMP, as for instance do the growth hormone, glucagon, and catecholamines, or inhibit it like insulin. The fat depots are reduced by starvation and built up by animals prior to hibernation for use during the abstention from feeding. Deposits of brown fat are formed before birth in man and many animals in the interscapular and axillary region and along the thymus and persist even in starving babies. Whether their size and composition varies with sex is not known.

The subcutaneous fat insulates the body against heat loss, is greater in women than men, and is of importance in temperature regulation (cf. below). The detrimental influence of obesity on life expectancy has been described for greedy mice, and is taken into account by life insurance companies in assessing the risk for the individual, and with it the amount of the premium. Fat is necessary, and non-esterified fatty acids are the main intermediate fuel for muscular activity in sheep. The link of obesity with other metabolic abnormalities is responsible for the adverse effect on the life span.

The skin is the most conspicuous feature in the appearance of mammals and is responsible for sex differences varying with species, as for instance in the length and density of the hair coat, the type of hair in various regions, the presence of horns and antlers, the colour of the skin and that of hairs, and the size and secretion of glands. Depending on the hair coat, the skin accounts for about 7% of the total weight as hide of cattle and for 12% as pelt and fleece of sheep. The skin glands vary in size from that of the large lactating breast and udders to that of sebaceous glands and the even smaller sweat glands. The secretion ranges from the production of milk to that of water plus salts, and to the specialized pheromones used as warning and attraction signals.

The skin of males is thicker than that of females and has more collagen arranged in larger bundles in the dermis, and around them a mesh of elastic fibres. These, in addition to the fluid held between collagen fibrils, and the muscle fibres, are responsible for the elasticity of the skin. Over-stretching by the accumulation of fat in the subcutaneous layer or in the abdomen by the enlarging uterus during pregnancy may be followed by the appearance of striae at the end of pregnancy or folds and creases after slimming. The collagen fibrils lose their water-holding capacity by coarsening with age, and lead to the wrinkles and crows' feet of elderly persons. The contraction of collagen bundles and inadequate regeneration of elastic fibres cause the persistence of scars.

The epidermis varies in thickness with the density and type of the hair coat, with sex, and in different parts of the body. The epidermis is thickest and has most keratin where there are no hairs, as on the soles, palms and pads, and is thinnest when there is a dense cover of fine hairs. The male rat has coarser and more widely spaced hairs and a thicker epidermis than the female. The face of men has a coarse coat of terminal hairs and a thick, keratinizing epidermis while in women a coat of fine down hairs protects a thinner epidermis. These differences are reflected also in the network of the superficial capillaries of the dermis, which is based on a thicker layer of fat and in combination with the fine hairs and the thin epidermis produce the complexion of young women, termed the 'English Rose'. The vascularity of the skin is a factor in its colour in addition to the amount of pigment. Teleangiectasy (a permanent dilatation of capillaries) causes the purplish nose of the heavy drinker. Vascularity and pigmentation are responsible for the high colour of sailors and farmers exposed occupationally to wind and sun. Vascularity causes the reddening and swelling of the sexual skin of monkeys, and in females varies with the sexual cycle. During the menstrual cycle of women ovulation raises the body temperature during the post-ovulatory (progestational) period and is accompanied by vasodilatation, increased secretory activity of the sweat and apocrine glands, and swelling of the breasts. The colour of the face of drills and mandrills is sexually dimorphic. In men and women the sun and genetic factors control the amount of pigmentation. The oestrogen surge of pregnant women leads to freckles in the face and pigmentation of the areola of the breast.

In mice, guinea-pigs and cats, the pigmentation of the hairs varies with sex when the colour gene is carried on the X-chromosome, and leads to mottling of the hair coat in females. The vellus hair of the pubic and axillary regions is replaced by terminal hairs in boys and girls, and in the face and on the chest in men. Androgens cause the reverse change from terminal to vellus hair in the alopecia of the scalp (baldness) of men and male monkeys, and this change can be induced in the females of *Macaca speciosa* by prolonged treatment with testosterone. It occurs also in some postmenopausal women, when the production of androgens in the adrenal increases relatively to the production of steroids by the ovaries. After puberty the hairs of the male rat are coarser and longer than in the female, and androgens render the hair coat of female rats and sheep coarser.

The breast and the apocrine and sebaceous glands of the skin are sexually dimorphic and controlled by the steroid sex hormones. The apocrine and sebaceous glands of the axilla and anogenital region in many species, the ventral pad glands of gerbils, and the flank glands of hamsters are larger in males than in females. The development of nipples in the mammary glands of male rats and lactation in male guinea-pigs can be induced by treatment with female sex hormones or antiandrogens in combination with prolactin. The glands, horns, and antlers change in the breeding season, and the hair coat alters in seasonal moults, at pregnancy and with the regular hair cycles, which differ with sex and are influenced by the gonads, the adrenals, the thyroid, the pineal body, the pituitary and the hypothalamus.

Differences in the proportion of internal organs.

The bones, muscles, fat and skin account for about two thirds of the body weight and only one third is made up of the central and peripheral nervous system, the heart with the blood-vascular apparatus and lympho-myeloid complex, the respiratory tract, the intestinal tract with liver and pancreas, the endocrine glands and the reproductive system. This section deals only with the gross differences in the weight of organs in relation to total weight of males and females, and leaves the analysis of the detailed sexual dimorphism in structure and functions of the internal organs for subsequent chapters. Some organs, such as the gonads and salivary glands of male rodents and boars, are more than twice as large as in females, and thus differ by a factor greater than that for the total body in size or weight even in rats and ferrets. They are thus larger not only in absolute terms, but also as a proportion of the total body. The absolute weight of an organ need not be proportionately larger if the males are larger than the females or vice versa. Thus the brain of men is heavier than that of women, but not in proportion to body weight of the two sexes. The adrenals in females are larger and heavier than in males, and also larger proportionately, as the females are smaller and lighter than males. The relation of absolute to relative proportions in males and females of the weight of bone, muscles and fat has been discussed in the previous part of this chapter.

The weight (strictly mass) of organs is the simplest measure for the comparison of sexual dimorphism, but is meaningful only when assessed as a proportion of the total weight of males and females, and the interpretation qualified. The gross weight of an organ includes the specific parenchyma as well as the supporting connective tissue, blood vessels, blood, nerves and deposits of fat, which are known to vary with sex; there is more connective tissue and less fat in male than in female organs, at least as a general rule. Some organs are sexually dimorphic in structure and function, as well as in overall size. The salivary glands of male rodents have a secretory tubule with a high epithelium, which produces a nerve and an epidermal growth factor, and at best is only rudimentary in females. The male kidney in some species is larger than the female organ, has a well-developed juxta-glomerular apparatus for the production of erythropoietin (necessary for the production of red blood corpuscles), a higher parietal epithelium in the Bowman's capsules of the glomeruli, and relatively more tubules and fewer glomeruli. Thus the gross comparison by weight, even in proportion to that of the body, does not necessarily compare like with like. That the sex differences in kidneys and salivary glands are real is proved by alterations following gonadectomy or the administration of steroid sex hormones, and implied by the pubertal and the menopausal changes. Comparison by weight does not allow for sex differences in the metabolic rates and the turnover times of parenchymal cells. An equality in these is assumed implicitly, but is largely unproven and is known not to apply in some organs; changes in the thyroid with the oestrous or menstrual cycle are known, and so are differences in the duration of the hair-cycle. On the other hand, the life span of erythrocytes is about the same in men, eunuchs and women whether prepubertal, mature or postmenopausal. For the majority of parenchymal cells hardly anything has been established about sex differences in turnover time, though some metabolic variations are recognized. In spite of all these qualifications, comparisons of variations with sex in the relative organ weights provide some useful information and indications of real distinctions.

The organs develop at different stages and rates, and their relative proportions to the body and to one another change with age. The central nervous system is the first to differentiate and to reach maturity, and loses its preponderance as other organs and systems differentiate and grow. Thus the relative weight of the brain decreases, and that of the more slowly developing locomotory apparatus increases with age. Sex differences alter with advancing age; the ovaries shrink after the menopause, while the testes retain their volume. There are very marked species differences in the sexual dimorphism of homologous organs, suggesting that they may play somewhat divergent roles

Table 6 *The sex ratios for the weight of the whole body of man, mouse and horse and the deviations in the ratios for the weight of some organs†*

	Man	Mouse	Horse
Whole body	1·17	1·02	1·06
Brain	−0·07	−0·09	−0·14
Lung	+0·03	0·00	−0·31
Heart	+0·11	+0·08	−0·05
Kidney	+0·13	+0·50	−0·32
Liver	−0·01	−0·15	−0·35
Spleen	−0·13	−0·23	+0·08
Adrenals	−0·20	−0·75	−0·35
Fat	−0·63	−0·20	− xx
Gonads	+4·43	+11·29	+ xx

†In Tables 6-8 the differences in the sex ratio of organs from that for the whole body are listed; for the human brain it is 1·10 and thus 0·07 lower than that of 1·17 for the whole body. A negative sign indicates a relatively larger organ in the female, a positive sign a relatively larger organ in the male. xx implies a reliable, though not quantitative estimate of a sex difference in excess of the sex ratio for total weight.

in the physiology. Table 6 compares the sex ratios for various organs in man, mouse and the horse. For the sake of simplification only the deviations from the sex ratio for the whole body are listed; for example, if the sex ratio for body weight is 1·17 and that for the brain 1·10, the difference of -0·07 is recorded. A negative sign indicates that the organ is relatively heavier in the female and a positive sign that it is heavier in the male as proportion of body weight. The figures quoted refer to young adults: humans aged 25 to 30 years, mice aged 24 weeks, horses of 2 to 3 years. The sex ratio for total body weight is greatest in man and smallest in mice. The females in the 3 species have relatively larger brains, livers, adrenals and more fat, while the testes are larger than the ovaries. For the other organs the sex ratios vary with species. The heart and kidneys are relatively larger and the spleen relatively smaller in male than in female humans and mice, while the lungs are equal or larger in the male. The mare has a larger heart and kidneys, but a smaller spleen than the stallion. Though accurate data relative to body weight are not readily available, the female pituitary, thyroid and thymus are larger in species with larger males and accepted as reliable estimates.

As the sex ratio for bones and muscles is usually greater than that for the total body in a species with relatively larger males, and as the blood circulation and respiration, served by the heart and lungs, are intimately involved with the functions of the locomotory apparatus, the sex ratio for these organs might be expected to vary in the same direction. While this is the case for man and mice, it does not hold for the horse. Similarly the kidney and the spleen play a role respectively in the formation and destruction of red blood corpuscles, and so should vary with the sex ratio for the whole body in the case of the human and murine kidney and against it in the spleen. The mare has both a larger kidney and a larger spleen. Unfortunately no adequate data are available for species such as the golden hamster or the rabbit to test whether the correlations in the sex ratios for the total body and those for the locomotory apparatus, heart, lung and kidney are reversed from those given for men and mice in table 6. There are probably exceptions in other species like those seen in the horse, which have to be explained by comparative physiological studies.

Table 7 *Stages in the development of the sex ratios for total weight and of the deviations from them of some organs in man†*

	Birth	10-11 years	17-18 years	20 years and over
Whole body	1·00	0·98	1·16	1·17
Brain	+0·02	+0·12	−0·06	−0·07
Lung	+0·01	−0·15	−0·04	+0·03
Heart	−0·05	−0·05	+0·04	+0·11
Kidney	0·00	−0·06	−0·08	+0·13
Liver	−0·01	−0·02	−0·11	−0·01
Spleen	+0·34	−0·02	−0·08	−0·13

† A negative sign indicates a relatively larger organ in the female, a positive one in the male.

Data about the relative proportions of organ to body weights at different stages of postnatal development are available for mice and men, though they are somewhat incomplete. They reflect, as might be expected, the differences in the onset and duration of the pubertal growth spurt and are shown in tables 7 and 8 for comparable phases: birth, childhood before puberty, the pubertal period and young adults. At birth, the sex ratio for the whole body in man (table 7) is 1·0 in the series analysed, though male babies tend to be larger than females. With the earlier onset of puberty, girls overtake boys at 10 to 11 years and are overtaken by them later on. There is thus a transient fall in the sex ratio for whole body weight, followed by a lasting increase. At birth, boys have relatively larger, though less mature, brains, lungs and spleens, and relatively smaller livers and hearts. All organs except the brain are relatively larger in girls when starting puberty at 10 to 11 years. The brain has stopped growing at that time, and the increase in size of other organs in girls, but not yet in boys, accounts for the different sex ratios. This is noticed particularly for the lung, kidney and spleen, and only the latter remains relatively larger. With male puberty proceeding, the other organs and the body outgrow the brain, and, since boys are larger than girls, the brain is proportionately smaller in them. The relative weights of the lung and heart become larger, but that of the kidney is slower in its increase. The table reflects very clearly the effect of the time difference in the pubertal growth spurt during the second decade.

Table 8 *Stages in the development of the sex ratios for total weight and of the deviations from them of some organs in mice.†*

	Birth	4 weeks	6 weeks	24 weeks
Whole body	1·01	1·01	1·04	1·02
Brain	−0·02	−0·05	−0·09	−0·09
Lung	−0·07	−0·04	−0·11	0·00
Heart	−0·01	−0·03	−0·08	+0·08
Kidney	+0·02	+0·02	+0·20	+0·50
Liver	−0·05	−0·07	+0·05	−0·15
Spleen	+0·10	−0·08	−0·24	−0·23
Adrenals	−0·16	−0·26	−0·53	−0·75

† A negative sign indicates a relatively larger organ in the female, a positive one in the male.

This difference is not seen in table 8 for mice, because it is much smaller. The sex ratio for whole-body weight is always greater than 1·0, with a transient rise at 6 weeks. At this stage the sex ratio for the brain decreases, and thus shows the same pattern as in man, while the liver shows a relative increase in the sex ratio which is subsequently reversed. The sex ratio for the spleen and adrenals shows a steep fall and that for the kidney a steep rise. The fall in the ratio for lungs and heart at this stage is subsequently reversed. The brain and adrenals are consistently larger and the kidneys smaller in females than in males. The ratio for the spleen alters in a similar manner in mice and men, but the development of the other organs differs markedly, and only a partial explanation is seen in the role of the growth spurt at puberty. Some of the pubertal changes take a long time for their manifestation and this is reflected in the changes in the sex ratios between 17 years and the young adults, and also those in mice between 6 and 24 weeks.

The sexes of probably all mammalian species differ not only in the configuration, function and structure of the reproductive organs, but in other features. The oestrous and menstrual cycle indicate the sexual dimorphism of the pineal-pituitary-hypothalamic centres and their feedback relations with the ovary, in contrast to the non-cyclic activity of the testes and their central regulations. The sex differences in size and configuration, in the proportion of the organs, and in their structure and function, vary with species. It is reasonable to assume in the absence of exact measurements, that female rabbits, golden hamsters, spotted hyenas and dwarf mongoose have relatively more bone and muscles than the males, since these body constituents usually account for some 40% of the total weight and these tissues are stimulated in their growth by the anabolic agents. The effect of the androgens in the males of these species must be less than that of the growth hormones in the females stimulated by the oestrogens acting on the size of the pituitary. There may be other factors involved in altering the hormonal balance and the metabolic rates and pathways.

These balances are fixed at different stages of development; the hypothalamic centres in rodents differentiate in the early postnatal period and respond to treatment with androgens or antiandrogens at that time, while this critical period falls into the gestation period of dogs and monkeys. Once differentiation has occurred, treatment with androgens, oestrogens and antiandrogens or by gonadectomy, does not alter the hypothalamic centres, though it interferes with the subordinate structures of the pituitary and the gonads and so with the feedback relations. The example of the percentage of bone, muscle and fat, and the onset of fattening as well as the growth rate in cattle, shows that steers retain some masculine features and have an intermediate rank between bulls and cows. These tendencies must have been fixed before birth either by hormonal or by chromosomal factors and are not altered by subsequent manipulations by hormones. How these balances are fixed in species with larger males and in those with larger females remains to be investigated.

The association in some species of a bigger locomotory apparatus in males with larger hearts, kidneys and lungs, and in others with relatively smaller ones (table 6), remains a problem of comparative physiology. Whether the deposition of fat in females in species with larger females is greater or less than that in the males can be found out easily by direct measurements. Other metabolic aspects of sexual dimorphism and their variations with species are more difficult to tackle.

10. Metabolism, thermoregulation and biorhythms

This section deals only with the general aspects of sexual dimorphism in the metabolism of mammals, and leaves the discussion of the specific features in the function and structure of organs to a subsequent chapter. The metabolic rate depends on the temperature within and outside the body, and in mammals some of the energy metabolism serves to preserve an even body temperature. The thermoregulation is thus intimately related to the general metabolism. The metabolic processes are controlled among other factors by the endocrine interactions and involve the sex hormones, which can influence some enzymatic activities. The endocrines in turn respond to external stimuli relayed by the sense organs to the cerebral centres, which coordinate them and regulate endocrine activities. Of the external stimuli the photoperiod is one of the most important factors in regulating the circadian cycles and the seasonal influences on the metabolism of mammals as well as the oestrous cycles, and is thus a facet in the sexual dimorphism of the energy metabolism.

The basal metabolic rate is measured in kilojoules per square meter of the body surface per hour ($kJ\ m^{-2}\ h^{-1}$) at rest and at a temperature demanding no special activity of the body such as shivering or sweating. The rate varies with species and within it with age and sex. The human basal metabolic rate rises from 126 $kJ\ m^{-2}\ h^{-1}$ at birth to a peak at 2 years of 239 in boys and 223 in girls, and then gradually declines to 147 and 139 respectively at the age of 70 to 74. The sex difference amounts to about 6% in the first decade, increases with puberty in the second decade to 10%, remains steady in the third and fourth and declines to 8% in the fifth and 6% in the eighth decade. The difference in basal metabolic rate is attributed to the greater amount of rather inert fat in the subcutaneous layer and throughout the female body. A similar relationship between the basal metabolic rate and fat content of the body, and so a sex difference, is found in cattle and rats. Muscular activity increases the energy requirements and with it the metabolic rate. At lower temperatures shivering increases the production of heat and at higher temperatures panting may dissipate heat and both these activities call for muscular action and thus an increase in metabolic rate.

Food serves to provide energy for the body after conversion into the constituents of proteins, carbohydrates, fats and innumerable specialized substances. The metabolism and choice of nutrients vary with the species; ruminants live on plants, carnivores on animals and omnivores use either. The choice of food is adapted to the differentiation of the digestive tract and is easily shown by the relationship between the length of the intestines and that of the body, which decreases from a factor of 27 for sheep and goats to 20 in cattle, 14 in pigs, 12 in horses, 6 in dogs, 5 in man and 4 in cats. Vegetarians thus have a long intestinal tract relative to the length of the body, while carnivores and omnivores have a relatively short one. This differentiation in relation to food has resulted in the evolution in cattle of a stomach consisting of a dorsal and ventral sac of the rumen, an omasum and an abomasum, of an enlarged colon in the pig, and of the caecum in the horse, and of a specialized flora and fauna inhabiting the digestive tract of predominantly vegetarian species which help in the conversion of the food. Though the food and its conversion into muscle differ in herbivores, carnivores and omnivores, the males of most of these species tend to be more efficient in producing meat than the females.

Feeding time as well as digestion vary with the type of food chosen. Cows graze for about 8 hours a day and spend an equal amount of time ruminating in approximately 4 to 7 periods of feeding and 15 periods of ruminating. The rest periods in cattle are broken up into multiple short intervals and are spent in sternal recumbency to allow the rumen to function. The circadian cycles of rest and activity are thus polyphasic in contrast with the monophasic cycle of activity and sleep in man, who spends less time eating and digesting than the vegetarian ruminant. The rumen is rarely empty, and the waste products appear in the faeces after 100 hours, compared with a bowel transit time of about 24 hours in man. The circadian rhythm of rest and activity has a greater amplitude in the monophasic than in the polyphasic species, where the multiple periods maintain a steadier and long drawn out metabolic activity. Sheep and cows close their eyes for at most 30 minutes a day, while man sleeps for about 8 hours.

There is a circadian rhythm in body temperature which affects the metabolic rate. The temperature of the core of the body measured in the rectum varies by 0·5 to 1 °C during the day, with maxima in the early afternoon and minima in the early morning in species with daylight activity, and the reverse for nocturnal animals. The range of oscillations in temperature over the day varies by 0·85 °C in men and 0·60 °C in women. The temperature varies with the menstrual period in women, and is higher in the second half of the cycle following ovulation than in the first half. The rectal temperature of humans is higher at night than the day and the reverse holds for the skin temperature. The body temperature results from the balance between the production of heat and its loss from the skin and in the breath, and varies with sex; women have cold hands but warm hearts as an old saying indicates. This is due to the gradient between the core and the shell temperature, with the female subcutaneous fat layer acting as insulator, helped by a difference in the capacity to sweat and in the management of the circulation of the blood. In hot humid surroundings women dissipate heat by a more rapid vasodilatation in the skin, while men sweat. Women sweat less and start the process later than the vasodilatation, which causes a relatively greater rise in body temperature and hence a lower relative heat tolerance. The capacity for sweating improves with repeated exposures as tested in cycling exercises, but does not reach the male level. The renal flow is reduced by the loss of water from the skin and in the breath, and also by the release of the antidiuretic hormone of the posterior pituitary. Premature babies are able to produce only little sweat, while normal babies sweat less than adults and lose heat by doubling the blood flow to the limbs and dissipate four times as much heat from the hands as adults. The sex hormones influence the sweat glands in development and function and the sex difference becomes manifest at puberty. This fact combined with the lower metabolic rate of women, the insulating effect of the subcutaneous fat layer, and the precocious vasodilatation in the skin, are responsible for the sexual dimorphism in thermoregulation. In hot surroundings, the blood is shifted from the viscera to the periphery, the heart rate is increased and the stroke volume reduced; this decreases the oxygen difference between the arterial and venous blood, which calls for increased respiratory activity and results in laboured breathing. The response to heat is by evaporation of water through the skin and respiration in man and cattle, and by panting only in dogs, pigs and rodents which are deficient in sweat glands.

In response to cooling, the blood flow in the skin is reduced to conserve heat, the hair coat is raised by the erector muscles (producing goose-pimples) to create an insulating layer of trapped air around the body, and the heat production is increased by shivering and by metabolizing the brown fat, which is not affected by starvation, but depleted by cold.

The reaction to changes in the ambient temperature is regulated in the brain stem in response to sensory stimuli from the periphery and the temperature of the blood around the hypothalamus. The injection of 5-hydroxytryptamine (5-HT), a neurotransmitter, into the ventricle of the brain of cats, dogs and monkeys raises the body temperature, but lowers it in rabbits, hamsters, rats and sheep, while nor-epinephrine, another neurotransmitter, has the reverse effect in the species named.

The photoperiod plays a role in the acclimatization to temperature; a short period of illumination causes deer mice (*Peromyscus maniculatus*) to adapt to cold conditions at 26 °C, and matches the effect produced by 6 °C during long periods of illumination. The heart rate of men is slower at rest during the night than at rest in a dark room during the day. For small animals the body temperature does not always vary with size of the body.

The dependence of the timing of the breeding seasons on the photoperiod in many species has been mentioned before. The duration of light and dark periods affects the oestrous cycles in rats and ferrets, and through it the metabolism of organs not directly involved with reproduction, and so produces a sexual dimorphism. The photoperiod obviously has a bearing on the metabolic rates in its circadian cycle, as for instance in nocturnal animals compared with species with daylight activity. Within the species some sex differences are found, and, though not fully explored, some illustrating examples have been recorded for small laboratory animals.

The extent of circadian variations in the reaction to injurious agents is demonstrated by the effect of a dose of endotoxin of *E. coli*, which kills 80% of mice when given in the early evening, and only 10% when given 8 hours later. Sex differences in the circadian rhythm of protein synthesis in some regions of the brain of rats (the thalamus) are shown by the rate of uptake of ^{35}S-labelled methionine, which reaches a peak value in males 8 hours earlier than in females. This timing is reversed in males castrated in the neonatal period, and in females androgenized at the same time by an injection of testosterone. Alteration of the lighting period is followed by an equal variation in the circadian rhythm, but this is not observed in young rats when the eyes are still closed, nor in castrate or spayed adults. This example shows an interaction of the perception of light, the photoperiod and the sex of the animal.

Fig. 9 illustrates the principle of the sex difference in the circadian rhythm. It assumes a 12-hour period of light between 0500 and 1700h, followed by an equal period of darkness, which is indicated by the horizontal bars at the bottom of the diagram. The sex difference in the timing of peaks and troughs of activity in the biological system is taken to be 12 hours. The maximal and minimal levels of activity are greater in this example in the female than in the male. While the

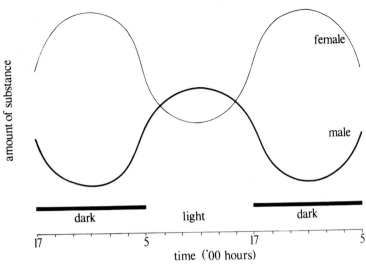

Fig.9. Sex differences in the circadian rhythm of metabolic activities.

timing and levels are imaginary, their pattern is not unlike that of protein synthesis in the brain of adrenalectomized rats. For other functions the symmetry of the waves may be compressed, with an unequal duration of rise and fall with sex differences in timing varying from 3 to more than 12 hours, so that there is an overlap in the amplitude of the waves. This may be ascribed to an increase in activity in the early or late phases of the light period in either sex. Thus variations in the details of this model are known to exist, but the principle remains the same.

The circadian and the oestrous cycle interact with one another. The activity of the thyroid gland in the rat, for instance, can be measured by the metabolism of labelled iodine in the blood. This is greatest in the morning of the day of oestrus, i.e. about 8 hours after ovulation, and is due to an increase of the level of the thyroid stimulating hormone (TSH) in the serum. This concentration varies, and except for the day preceding oestrus (the day of prooestrus) reaches a peak at 1200h and a trough at 2100h when the animals are kept at a temperature of $25 \pm 2\,°C$ and in light between 0500 and 1700h. On the day of prooestrus, however, the peak is higher and occurs at 1700 and the trough at 0900h. In the male rat the level of TSH in the serum is higher than in the female, but the circadian rhythm has a similar pattern except for the change in the female at prooestrus.

In fig. 10, the sex difference in respect to the circadian rhythm is depicted in a diagrammatic form which bears some resemblance to the

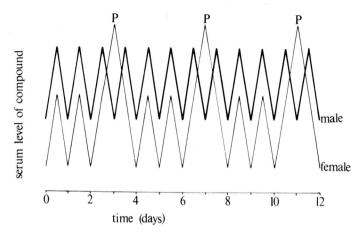

Fig. 10. Interaction of the circadian and oestrous cycle. The oestrous cycle is assumed to last four days and P denotes the day of prooestrus.

pattern reported for the TSH levels in the serum of rats. The level of the assumed compound is greater in the male than the female, but the pattern of rise and fall is similar except on the day of prooestrus (P), when the peak in the female is higher than that in the male. A 12-hour rhythm for the photoperiod and for the circadian rhythm is assumed, and a duration of 4 days for the oestrous cycle. Thus the maxima and minima of the concentration occur at the same time on the first, second and fourth day of the oestrous cycle in both sexes, but on the third day a peak level in the female coincides with a trough in the male. The diagram does not aim at accuracy as regards the timing of the circadian and oestrous cycles, and is meant to illustrate in principle the superimposition of the oestrous on the circadian cycle by a surge in the release of a compound at a specific interval. The surge of TSH at prooestrus is accompanied by similar rises in LH (luteinizing hormone) and PRL (prolactin) though there are differences in the height and timing of the oscillations.

Reversal of the photoperiod by 12 hours leads to an alteration of the timing of the oestrous cycle by the same amount. In normal light-dark sequence there is a critical period for the initiation of ovulation between 1400 and 1600 hours on the day of prooestrus. An injection of a barbiturate anaesthetic (Nembutal) given at that time, and at that time only, delays ovulation by 24 hours, and the same effect is obtained if the treatment is repeated after 24 hours. The circadian cycle can be varied by varying the photoperiod, and altered by events in the oestrus cycle, which in itself is subject to the photoperiod and to other endocrine influences. Thus neither is immutable, and neonatal treatment of females with androgen and of males with antiandrogens can suppress or induce such cycles in the hypothalamus. The absence of oestrous and menstrual cycles in the males is one of the important sex differences affecting reproductive organs, as well as those not directly concerned with propagation.

The relation of the photoperiod to cerebral centres concerned with the production and release of neuroendocrine secretions, and with their communication via neurotransmitter substances, is illustrated by some experiments on Syrian hamsters. If males are kept at constant temperature and continuous illumination between 0500 and 1700h and are injected at 0900h with melatonin, a hormone produced by the pineal body, their testes decrease in mass. An injection given at other times does not have this effect. In females kept under the same conditions of lighting and temperature a dose of melatonin given at 1500h inhibits the oestrous cycle, but the same dose given at 0800h does not, nor is it effective after pinealectomy. The concentration of serotonin (5-hydroxytryptamine = 5-HT), the precursor of melatonin, has a marked circadian rhythm in the pineal gland, with a peak at 1300h and a trough at 2300h. This rhythm persists in rats kept in the dark or blinded, but is abolished by continuous illumination and also by the removal of the supracervical sympathetic ganglion.

The concentration of corticosteroids in the plasma of adult rats varies with the lighting periods and with the oestrous cycle. Constant light induces persistent oestrus after about three weeks, but affects the circadian rhythm within three days. Mice kept continuously in the dark retain a diurnal rhythm of adrenal activity, with a peak at around noon and a trough at midnight. This independence from a regular photoperiod is described as a 'free-running period' and apart from the adrenal activity in mice applies also to their behaviour on running wheels, the movements of which can be recorded continuously. Male mice kept in darkness for nine months are active at regular intervals. If such mice are castrated, and one group subsequently implanted with pellets of testosterone—operations which require exposure to light on three occasions for two hours each— the activity is found to be decreased by castration and restored by the insertion of the testosterone pellets. In the free-running periods, the rhythm of circadian activity is thus still subject to the influences of sex hormones.

The adrenal gland of rats, isolated and grown in culture outside the organism, retains the circadian rhythm of steroid production in the absence of the pituitary adrenocorticotropin (ACTH) that normally regulates the activity. The cultures are kept in the dark except for the short periods required for a change of the nutrient medium. Thus local factors can act as biological clocks in isolated organs without a regular photoperiod or central control. There is good evidence for circadian rhythms in the regulating centres such as the pineal gland and the hypothalamus, as well as in the subservient pituitary. These are affected by the feed-back from the periphery; in the example of TSH level in the serum of rats and its rhythm, the sex difference in the amount of the substance present is due to the damping-down of the pituitary output by the greater activity of the thyroid gland in females. Nor are the circadian rhythms of different organs synchronous. The mitotic activity in the murine adrenal is at peak values at noon, but that in the epidermis at night.

This is also shown in the human serum levels of various hormones; the growth hormone reaches a maximum about one to four hours after the onset of sleep, prolactin in men at 0500h and in women a few hours earlier, ACTH on waking followed by cortisol one to two hours later, and testosterone at 0800h. In the rat the timing appears to be somewhat different, with peaks for cortisol reported at noon by some investigators and in the evening by others, while prolactin, growth hormone and TSH have maxima at about 1100h. The level of growth hormone varies in response to insulin levels, and these with the glucose level in the blood, which depends on the intake of food. Eating stimulates the activities of the intestinal tract including that of the gastro-intestinal hormones (secretin, gastrin, pancreozymin and enteroglucagon), which affect the pancreas and thus the output of insulin. To some extent the differences in the timing of the circadian rhythm in man and rats may reflect the daylight and nocturnal activities of the two species. The response to the stresses of light may be similar, since darkness does not abolish the circadian rhythms in rats or blind people.

The metabolic circadian activities are obviously more pronounced in monophasic than in polyphasic species and milk production in ewes and cows is not, or not greatly, affected by the photoperiod. Breeding activities, however, vary in these species with the photoperiod circannually. With daylight reduced from 16 to 8 hours the concentration of LH in the pituitary of rams is doubled and that of FSH quadrupled, illustrating the role of the photoperiod for the initiation of the breeding season. The internal clock does not ensure synchrony of activities in all organs, and its working varies with species. Some of the oscillations are due to the time scale of metabolic activities in the peripheral organs; DNA synthesis has a fixed minimal period and ordered sequence of events, and so have many other synthetic processes depending on the availability of precursor substances from other organs or activities. These time-consuming steps, once initiated, may be responsible for the persistence of circadian rhythms in the activities of explanted organs, and for the free-running periods of animals in conditions of constant temperature and illumination. In the latter circumstances the feed-back relationships between the various members of the endocrine system and the centres may still function, though in a modified form.

The brain plays the major role in coordinating the oscillations during the daily, seasonal and other periods, and attempts to locate the coordinating centres are being very actively pursued. Such governing regions must integrate the environmental factors such as light, temperature, weak magnetic, or electrostatic or γ-ray fields which act as synchronizers of the biorhythms with the endogenous oscillations of the internal clocks. Damage to some parts of the brain interrupts the cycles, and points to the involvement not only of the neuroendocrine regions, but of the neurotransmitters and the sensory connections with the peripheral sense organs. One of the suspected pathways leads from the retina (via the inferior accessory optic tract to the superior cervical ganglion of the sympathetic nervous system, and from there via sympathetic nerve fibres) to the pineal gland, which is linked with the hypothalamus and other cerebral regions. The pineal gland contains an enzyme (HIOMT = hydroxy-indol-O-methyltransferase) which converts the neurotransmitter serotonin into melatonin. This conversion has a circadian rhythm in the rat pineal gland which persists in darkness and in blind animals, but is abolished by continuous illumination or by transection of the accessory optic tract, or by excision of the superior cervical ganglion. The pineal gland has a high concentration of the neurotransmitters 5-HT and the dopamine (DA) derived nor-epinephrine. The 5-HT is present in parenchymal cells and in the dense core granules of the numerous sympathetic nerves present in the organ. The latter contain also the nor-epinephrine, which is formed from tyrosine via L-dopa and dopamine. It reaches a high level in the pineal body at the end of the dark period and falls to a low level at the end of the light period, and this rhythm is abolished by the deprivation of optic stimulation. There exists some antagonism between 5-HT and nor-epinephrine, since the increase of one substance leads to a decrease in the other. In isolated pineal glands the conversion of tryptophane via 5-hydroxy-tryptophane to 5-HT and melatonin is increased by the addition of nor-epinephrine, and thus the concentration of 5-HT is reduced.

An alternative pathway leads from the retina via the retinohypothalamic tract to the suprachiasmatic nucleus which receives serotonergic fibres from other parts of the brain, and whose cells contain the gonadotropic releasing hormone (LRF), which regulates some functions of the hypothalamic-pituitary system. The concentration of 5-HT in the suprachiasmatic nucleus is greater in the female than the male after puberty. These attempts at locating the centres regulating the biorhythms are very tentative, and are likely to be modified by further investigations. Initially they have been considered in terms of the feed-back mechanism between the peripheral endocrines and the pituitary, and subsequently the pituitary-hypothalamic system has been added. These regulators are connected through the neurotransmitters with other parts of the brain and the sense organs, and through them as well as the sensory nervous systems to the target organs. A number of substances are involved in this chain of interlinking processes; the dopamine-serotonin transmitters serve the communication between nervous elements in the brain, spinal cord, spinal and sympathetic ganglia. To link these structures and the receptors in the target organs cholinergic and adrenergic nerves are used, which release acetylcholine and adrenaline respectively. Nor do the coordinating substances end at this point, since within the target organs the individual cells have to be regulated in time to act synchronously in the production of various substances, and for that purpose the short-distance communicating substances of the APUD system and the prostaglandins are used. How many of these functions show a sexual dimorphism has so far been established in only a few examples, such as the oestrous and menstrual cycles.

The sex differences in biorhythm deserve attention not only for their theoretical importance, but also for their practical implications in the timing of therapeutic interferences and the choice of the right dose. Biophysical factors modify the effect of drugs; alcohol is rapidly absorbed from an empty stomach, and quickly reaches a fairly high concentration in the bloodstream, and is thus more intoxicating than when taken during or after a meal. Agents aimed at inhibiting or destroying cancer cells, such as irradiations or carcinostatic drugs, are likely to be more effective if applied in time for a sensitive stage of, for instance, the process of cell division, which may be that of DNA synthesis and so the S phase or the period immediately preceding M (G2 in fig. 1). Investigations on the best timing in the fractionation and size of doses of irradiation have been proceeding for the last 50 years. The differential effect of endotoxin on mice depending on the circadian rhythm, and the critical period for the use of anaesthetics in delaying the LH surge and with it of ovulation in rats, have been described above.

Sex differences in the sensitivity to drugs can be elucidated only in large scale trials, and are thus rarely possible in people needing such substances for distinct therapeutic purposes. Unfortunately, findings on animals cannot be extrapolated to other species and man without qualifications. After puberty a male rat requires four times the dose of hexobarbitol for the same anaesthetic effect as the female matched for size and age, because he has four times the enzyme activity in the liver for metabolizing the anaesthetic. There is no sex difference in the response of mice and guinea-pigs to this barbiturate. Female rats increase their enzymatic activity in the liver on being given testosterone, and thus are more tolerant of the effect of the anaesthetic, while oestrogens have the reverse effect in male rats. A number of drugs are more easily tolerated by males than females, and also vice versa. Male rats, for instance, are more resistant to nicotine, picrotoxine, sulfanilamide, DDT and barbiturates, females to morphine, cocaine, ergot and ethylene glycol. Male mice of some strains are exceedingly sensitive to chloroform vapour, while females are unaffected by it. Women are reported to break down aspirin more slowly than men, and thus are more likely to suffer from the effect of the medicament on the gastric and intestinal lining. These differences in response to drugs are indicative of sex differences in metabolism, and so are the relative frequencies of metabolic disorders, e.g. of diabetes and gout, but for only few drugs has the influence of biorhythms, such as the circadian variations and the modifying effects of the oestrous or menstrual cycle, been determined. Since the metabolic rate is affected, it is more than likely that the metabolism of therapeutic agents and poisons will vary as well. For example, the activity of the liver enzyme tyrosine transaminase in the rat has a circadian cycle which persists in the dark, but is abolished by continuous illumination; the activity is increased by a reduction of the level of norepinephrine in the brain by, for instance, the drug reserpine, and reduced by raising the level of the neurotransmitter in the brain through agents which block its degradation. The hydroxylase activity in the liver of male rats is greater than that in females, and varies during the day. Adrenalectomy of male rats reduces the metabolism of hexobarbitol by liver microsomes *in vitro*, but has no effect on females. Thyroxin inhibits this metabolism in males, but not in females.

The sex differences in metabolic rates and thermoregulation are dependent on those of the enzyme activities and other functions of the peripheral organs, on their variations with biorhythms, on their sensitivity to the sex hormones, on the cholinergic and adrenergic innervation, on the coordination of the cellular activities by the APUD and prostaglandin system, on the feed-back to the neuroendocrines and the central nervous system, on the central neurotransmitters and their regulation by the coordinating and interlocking cerebral centres and the input of external stimuli.

11. The blood-vascular system and the lymphomyeloid complex

Apart from their own specific contributions, these two systems serve by their corpuscular and humoral components to transport the secretions of other organs, nutrients, oxygen etc. throughout the body. Both the cells and the composition of the blood plasma and of the lymph vary in detail with sex. Variations in the levels of various hormones in the serum have been mentioned and those in other substances are too numerous to catalogue here.

The sex differences in the effective size of the bone marrow and the lymphatic organs, which produce the cellular elements of the blood and lymph, are difficult to estimate. These organs have a resident as well as a transitory cell population, quite apart from the blood vessels and connective tissues. Data for the volume and weight of the bone marrow in males and females are not adequate, and the relative amounts have to be guessed from the number of cells produced and other functional differences. The spleen is proportionately larger in the males of some species (horse, table 6) and the females of others (man and mice, tables 6 to 8). The thymus of rodents is considerably larger in females than in males, and contains more secretory cysts and tubules, which vary in size and activity with the oestrous cycle. In female guinea-pigs the proportional weight of the lymph nodes and lympho-epithelial organs is greater than in males. Since all these tissues respond differentially, with increases or decreases in volume and activity, to the steroid sex hormones and castration or spaying, the sex differences can be accepted as real.

The influence of sex on the cells produced by the haemopoietic tissues varies with species. One millilitre of blood contains more red blood cells in men, bulls and stallions, and in male beagles, cats, rats than in females of their species, while the number is equal in sheep, goats and rabbits of either sex. In normal individuals, the haemoglobin level of the blood varies with the erythrocyte count and is thus greater in the males of some species. Castration of horses, pigs, cattle and hamsters reduces the count of red blood cells per millilitre to a lower level than in females. The amount of erythropoietin which is necessary for the production of haemoglobin is greater after puberty in men than in women and the same applies to males of other mammals. Androgens as well as a variety of other substances (cobalt, TSH, ACTH) stimulate its production in the kidneys, while oestrogens inhibit it. The juxtaglomerular apparatus, composed of specific cells of the epithelium surrounding the glomeruli, is believed to synthesize the substance, though in rats and dogs it does not respond with apparent changes to the administration of testosterone, which induces a hypertrophy of the kidneys and raises the level of erythropoietin.

The lower count of red blood corpuscles and level of haemoglobin in women is not due to the menstrual loss of blood, since it persists after the menopause and is seen also in mammals that do not lose blood during the sexual cycles. Administration of androgens to females of some species induces a polycythaemia. Sex differences in the diameter and life span of erythrocytes are not consistent. Though their life span has been estimated as 120 days in men and as 109 in women, the assessments differ with the method used.

Blood platelets are formed by the megakaryocytes of the bone marrow and spleen, where a third of them are stored, while the rest circulate in the blood stream surviving for 8 to 12 days. The level decreases by about 20% during the first half of the menstrual cycle, and is lowered also by the administration of oestrogens. In dogs, oestrogens deplete the megakaryocytes and depress the level of platelets, while androgens do not appear to have any effect on them.

The count of granulocytes (neutrophile, eosinophile and basophile leukocytes) varies with the time of day, the intake of food and the presence of even minor infections. The neutrophile count is low in the morning, higher in the afternoon and at both times greater in women than in men. Similarly male mice have a lower count than females. The life span of granulocytes is estimated at 4 to 5 days in the tissues, and less in the circulation. In bone marrow transplants, red blood cells survive for approximately 100 days, platelets for about 9 days, and granulocytes have a half-life of around 6 hours.

The lymphocytic system and its role in the immune defences of the body have attracted considerable attention in recent years, and a variety of cell populations with diverse functions have been recognized. Two main groups are distinguished: T-cells (thymus-dependent) and B-cells which synthesize antibodies. Neither group is uniform, each being composed of many different clones of specialized cells. In man, about a quarter of the circulating lymphocytes are B-cells and three quarters are T-cells. Their role in immune reactions to foreign material, whether living or not, differs: the T-cells are responsible for cell-mediated and the B-cells for humoral immune reactions. The

T-cells are made competent by factors imprinted on them in the thymus; they attack foreign cells and microorganisms, are cytotoxic, and stimulate mononuclear macrophages. They are diverse in their responses, and have a memory. They alone respond to photohaemagglutinin or concanavalin with DNA synthesis and enlargement to blast-cells, and are found in the deep cortical regions of lymph nodes and in the parafollicular and perivascular areas of the spleen. Some of them are short-lived, while others survive for about 80 days in mice and 5 to 10 years in man.

The B-cells produce a variety of immunoglobulins called A, D, E, G and M and abbreviated to IgA, IgD, IgE, IgG and IgM, and are more sessile, and inhabit the outer cortical and medullary regions of lymph nodes, Peyer's patches and tonsils. They can be identified by immunofluorescent markers on their surface, and enlarge to form plasma cells. The lymphocytes, like all blood cells, originate in the foetal yolk-sac, pass through the liver to the bone marrow, and from there to the lymphatic tissues in the lymph nodes, lymphoepithelial organs, thymus and spleen. The T-cells are recognized by the formation of rosettes around foreign cells such as sheep red blood cells. A strain of mutant mice ('nude' mice) has no thymus, and there are thus no T-cells, though the B-cells are present. Such mice, like others thymectomized in the neonatal period, are unable to reject heterografts and are useful in transplantation experiments that are difficult or impossible to perform in normal animals.

Though there is no direct evidence for sexual dimorphism in the proportion of B- and T-cells, it is not unlikely since the size and structure of the thymus, of the spleen and of the lymphatic tissues varies with sex, and so do the immune responses and the levels of at least one of the immunoglobulins. The spleen is larger in stallions than in mares, and in women and female mice. The female thymus is larger in most mammals examined, and the lymphoid complex larger in female guinea-pigs. Castration of immature mice causes hypertrophy of the thymus, spleen and lymph nodes. Simultaneous thymectomy prevents the enlargement of the lymph nodes. Splenectomy has no consistent effects on the population of cells in the bone marrow, in the lymphoid tissues and in the circulation, but it raises the sex ratio in litters of mice immunized with skin grafts or spleen cells from males.

As the structure and cellular composition of the spleen and lymph nodes do not appear to vary with sex, an increase in weight of these organs may suggest a uniform increase in all components. The thymus of rodents is sexually dimorphic in the lymphoid and epithelial elements of the cortex, but not in the medulla, the Hassal's corpuscles or the parathyroid-like formations. The cortical and medullary epithelia differ morphologically and in their reaction to gonadectomy and to steroid hormones. Castration induces a hypertrophy of the cortex, and the involution of the cortex in males starts at puberty. Testosterone reduces the weight of the hypertrophied organ in castrate rats, while oestrogens have only a minor effect on the thymus of spayed animals. Regeneration of the cortex following exposure to X-rays or treatment with hydrocortisone is faster in females than in males and retarded in the latter by androgens.

The secretory ducts and cysts are more numerous in female than male rats, and vary in size and activity with the oestrous cycle. Oestrogens increase their incidence and size in males, while androgens reduce it in females. The functional role of these structures is not known. They may or may not be involved in the formation of 'thymosin', a thymic extract which, injected into neonatally thymectomized mice, increases the population of lymphocytes and the immune reactions against skin transplants. Prepubertal castration of mice enhances the cell-mediated immune response to skin grafts, and the rosette formation against sheep red blood cells, but does not affect the production of antibodies to foreign serum. There is thus a distinct link between the sex difference in the thymus and the response of T-cells to immunological challenge.

The activity of some of the B-cells is correlated with the number of X-chromosomes in an individual, since the serum level of IgM is higher in people with an XXX or XXXY constitution than in those with XX or XXY, and is lowest in the normal male with the XY constitution. The IgM level depends on which X chromosome is transmitted, as it is closer in mothers and sons than in mothers and daughters, and closer between fathers and daughters than between fathers and sons (the sons receive the maternal X-chromosome and the daughters both a paternal and maternal X-chromosome). Strains of mice differ in the level of IgM and thus in their response to challenge by a pneumococcal polysaccharide; the BALB mice are high-responders while the CBA mice are low. Crosses of a low-responding female with a high-responding male result in sons with a low response carrying the maternal X-chromosome, while the daughters having a maternal and a paternal X-chromosome have a response intermediate between high and low. A back-cross of such females with high-responding males results in daughters all of whom are high-responders, while only a few of the sons are. This correlation with the X-chromosome holds only for IgM and not for IgA or IgG, which have a higher level in the serum and a longer half-life.

The immune competence of female mice is greater than in males, and is enhanced in the latter by oestrogens. A suggestion of sexual dimorphism in human immunological competence is provided by the prevalence of autoimmune diseases in women and immune-deficiency in men. In autoimmune diseases antibodies are formed against the person's own constituents which, following an injury to cells, are shed into the circulation. The best known example is a condition named Hashimoto's thyroiditis, in which autoantibodies against the thyroid affect the organ. In persons with rheumatoid arthritis a 'rheumatoid factor' is found which belongs to the same class of substances and is present also in 5% of the normal population. These autoantibodies tend to increase in their prevalence and titre with the age group of the population. In immune-deficiency states the production of immunoglobulins is deficient; boys are mainly affected and unable to cope with even minor infections.

The subject of transplantation of tissues and organs with the rather complex conditions for tolerance and rejection is beyond the scope of this monograph, except for the examples of sex-linked acceptance and rejection of skin grafts between siblings and histocompatibility factors on either the X- or Y-chromosome. An instance of an X-linked histocompatibility is provided by experiments involving reciprocal crosses between two strains of mice (C57Bl and BALB) in which a male from one strain is crossed with a female of the other and vice versa. The offspring in the first filial generation (F1) may thus have a father either from the C57Bl strain and will be called CF1, or from the BALB strain and will be called BF1. CF1 males serve as hosts for skin grafts from the other combinations. If the transplants stem from other CF1 males they are accepted, but those from BF1 males are not. The tissues differ in the derivation of their X-chromosome, which in the CF1 male comes from the B mother and in the BF1 male from the C mother. Random heterochromatization of the B and C-derived X-chromosomes leads to rejection of half the tissues coming from a CF1 female and the acceptance of almost all those from a BF1 female. Thus the presence of an active B-derived X-chromosome is necessary for acceptance by the CF1 male host; the B-derived X is present in CF1 males, absent from BF1 males, present in half the cells of CF1 females and in all BF1 females.

A Y-linked histocompatibility factor can be demonstrated in a substrain of C57Bl mice (C57Bl/6), where the skin grafts from females are rejected by males and tolerance is not increased by injection of testosterone into the females or by castration or by oestrogen treatment of the males. Female hosts will accept grafts from donors of the same strain with an XO constitution, but not from those with an XXY constitution. Females can be made tolerant, however, by hyperimmunisation with very large doses of spleen cells from males or of skin grafts from newly born males, while female spleen cells are not effective in increasing the tolerance. The age of the host, but not of the donor, is of importance in the success of skin grafts from males to females of the same strain—in this instance strain A. The percentage of takes decreases from 75% when the hosts are 5 weeks old to nil when they are 39 weeks.

The age factor in the change of immune responses in mice resembles the increase with age of autoimmune factors in women. This change is probably not closely related to the involution of the thymus, which usually starts soon after puberty. The secretory activity of the thymus starts shortly before birth, and coincides with the stimulation of lymphopoiesis. It subsequently wanes and varies with sex and in response to sex hormones. The influence of the age, sex and hormonal factors on the main types of lymphocytes and the many subgroups or clones of lymphocytes is as yet incompletely understood. Nor is there adequate information about the quantitative relationships between the various types of lymphocytes and their products. Small numbers of foreign or ectopic cells and their products induce an immune reaction, while large numbers may suppress it. Similarly the tolerance of transplanted tissues varies with organ; it is good for corneal grafts and less so for transplants of liver, kidney, heart, bone marrow and skin. Yet females fertilized by sperm from related males or those of the same species still reject skin grafts from closely related males.

Some of the sex differences in congenital abnormalities of the heart and of the vascular system have been described above. The blood pressure rises fast in infants during the first 2 years and then more gradually until puberty, when it increases more in males than in females. The level reaches a plateau in the third and fourth decades, and then rises again in men, while in women it starts to rise with the menopause. Abnormalities in blood pressure, such as essential hypertension, are more frequent in men up to the age of 40 years, and occur in some families, which suggests a congenital component. It is due to the stenosis of renal arteries which alters the feed-back relations of renin-angiotensin and aldosterone secretion. Men of some families are predisposed to coronary atherosclerosis. This condition may be due to a genetic background as well as to higher serum levels of β-lipoproteins and cholesterol in men than in premenopausal women and castrates. Young women have more α-lipoproteins than menopausal women or men, and oestrogens may change the pattern of lipoproteins in postmenopausal women to that of younger females. These observations point to sex differences in the various localizations of the vascular system and in particular of the arteries, and also in their sensitivity to the content in type and amount of lipoproteins.

12. The endocrine system

The endocrine system comprises not only the major glands (pituitary, pineal body, thyroid, parathyroids, pancreas, adrenals and gonads), but also the very numerous components of the intestinal tract, kidney, thymus and salivary glands that have some endocrine function. The centres in the brain stem govern the activity of the whole system with a series of feedbacks and interaction between the glands and in collaboration with the nervous system. The whole complex represents the integrating force for the coordination of cellular activities within organs, for that of organs and all metabolic processes, and for growth, differentiation and other functions. The range of communication varies from those hormones affecting merely neighbouring cells, as do the peptides with paracrine functions, through those acting on adjacent organs as do the gastrin, secretin, cholecystokinin and enteroglucagon produced in the intestine and acting on the pancreas, liver and gall bladder, to those regulating the activities of a variety of distant organs directly or via their effect on the pituitary and hypothalamus, such as thyroxin, parathormone, calcitonin, insulin, adrenalin, corticosteroids and gonadal hormones. An increase in the circulating thyroid hormone not only depresses the secretion of TSH, but promotes that of ACTH, prolactin and growth hormone and stimulates metabolic activity in the periphery. Oestrogens affect the pituitary production of the gonadotropins, prolactin, TSH, ACTH and growth hormone by stimulating growth of the pituitary, and affect directly the activity of the adrenal cortex and thyroid. The direct and indirect actions of the gonadal hormones form the basis of the sex differences in the endocrine system.

The target for the endocrine factors may be restricted to receptors in some organs, as for instance the genital tract, or may be distributed all over the body, as is the case for the growth hormone and its active derivatives. The secretions may belong chemically to the steroid series (corticosteroids, testosterone, oestrogens, progesterone), may be peptides (APUD series), glycoproteins (pituitary hormones) or proteins (insulin), or may be derived from fatty acids (prostaglandins). While sex differences in the activities of the major endocrines have been fully documented, those for the diffusely distributed secretory structures have not been sufficiently investigated. The exceptions are the production of erythropoietin or its precursor in the juxta-glomerular apparatus of the kidney, and that of the epidermal and nerve growth factors in the ducts of the submaxillary gland, both of which are secreted in greater amounts in the male than in the female.

The dilution in the concentration of hormones released into the blood increases with the distance from the source. Hence some glands are close to the principal target organs, as for instance the intestinal endocrine cells to the pancreas, or linked with it through a portal venous system, as are the intestines and pancreas with the liver and the hypothalamus with the pituitary. In both examples the blood is collected from the capillaries into veins, which do not empty it into larger veins, but on entering the target organ form another capillary system, and thus deliver the transported substances with very little dilution. Many of the hormones are bound to proteins before or after their release into the blood, and are thus inactivated until they reach a receptor site which can reactivate them. Thyroxin and testosterone are examples of this phenomenon. Most of the larger endocrines produce more than a single active agent, and these may not necessarily be related to one another in their function; the thyroid secretes thyroxin as well as calcitonin, the adrenals produce the corticosteroids and the epinephrine series. In others hormones with opposite effects may be produced; the pancreatic islets secrete insulin to lower, and glucagon to raise the level of blood sugar; the ovaries produce oestrogens and progesterone as well as testosterone; the pituitary has to manufacture a great series of compounds with regulatory activity. The major endocrines are thus not receivers and senders of single signals, but receive a variety of messages and in their turn transmit them to the periphery, where they are processed again. The superior centres in the hypothalamus integrate external stimuli as well as those from the inside of the organism including the feed-back from the endocrines; being sexually dimorphic they impose differences on the activity of the endocrines as well as the circadian, menstrual and oestrous rhythms or their absence.

The pituitary of most mammals is larger in the female than in the male in proportion to the size of the body, and is enlarged during pregnancy and lactation. Oestrogens stimulate growth of the pituitary and thus promote the production of growth hormone, which is in turn reflected in increased growth of the body. The hormones are produced in the anterior pituitary (adenohypophysis) in specialized cells called after their product thyrotopes (TSH), gonadotropes (FSH and LH) and so on for ACTH, growth hormone, prolactin, melanocyte stimulating hormone (MSH). These cells synthesize specific substances and in this respect differ from one another. All of them may be stimulated simultaneously in their rate of synthesis and release of their specific products. Oestrogens affect not only the gonadotropes, but also the cells producing TSH, ACTH, prolactin and growth hormones and similar actions are known for other hormones. This feed-back in the regulation of pituitary activity may be direct from the peripheral endocrine or mediated via the hypothalamic centres. This is illustrated dramatically by gonadectomy, which in the absence of the damping down from the peripheral endocrine secretion results in a hyperplasia of the gonadotropes in an attempt at compensatory stimulation of the gonads and subsequent exhaustion of synthetic granulations which characterizes the 'castration cells'. Pituitary adenomas composed of specialized and often functioning cells are thus produced, and can be propagated in culture *in vitro* and used for obtaining TSH, gonadotropins, ACTH and prolactin. Similarly ovarian tumours are induced by placing the ovaries into organs draining through the portal vein into the liver, where the oestrogens are metabolized before reaching the general circulation. Thus the pituitary and hypothalamus are fooled into producing gonadotropins which stimulate the growth of ovarian tissues resulting in tumour formation there, and subsequently in the pituitary. Both ovaries have to be removed from their normal location and prevented from discharging into the general circulation to obtain this result, otherwise the oestrogen reaching the brain before the liver will signal that no excess stimulation is needed.

Apart from the periodic variation in the output of FSH, LH, ACTH, TSH and prolactin with the oestrous and menstrual cycles, the average levels of prolactin are higher in the female, and those of TSH in the males of rats and in women of reproductive age. The pituitary is also differentially sensitive to the administration of oestrogens and androgens; small doses of oestrogens inhibit the secretion of LH, which is necessary for maintaining the function of interstitial cells in the male and female gonads, but androgens are less effective in this respect. Large amounts of oestrogens suppress the secretion of pituitary gonadotropins, and so cause atrophy of male and female gonads. Unilateral gonadectomy in rats causes compensatory hypertrophy of the remaining gonad of unequal extent; the weight of the remaining ovary increases by 40%, that of the remaining testis by only 12%. The pituitary secretion of gonadotropins is more rigidly controlled in females than in males; excessive amounts are produced by ectopic pituitaries derived from females, but not in those of male origin. The testicular functions are maintained after hypophysectomy of rats by LH alone and following gonadectomy the levels of LH are increased more in males than in females, while for FSH the reverse holds true. The possible differences in the hypothalamic control of gonadotropins will be discussed in the next section.

The posterior pituitary (neurohypophysis) produces or stores the antidiuretic hormone (ADH), also called vasopressin, and oxytocin. ADH regulates the water balance by increasing the permeability to water of the distal portions of the renal tubules. Oxytocin stimulates the activity of the myoepithelial cells of the mammary glands, which help in the ejection of milk, and also uterine contractility. Sucking of the nipples or stimulation of the peripheral receptors in the genital tract initiates the release of oxytocin into the circulation, indicating that the peripheral stimuli are transmitted via the nerve pathways and the spinal cord to the midbrain and thus to the neurohypophysis. Though produced in the male, the function of oxytocin in males is unknown.

Sex differences of the pineal body are not fully documented. The secretion of melatonin is implicated in the control of gonadotropins, of the circadian rhythms, of the onset of breeding seasons and of puberty, as well as in the maintenance of gonadal functions, but the effect of sex and of sex hormones on the structure and function of the gland are not fully established. It is known, however, that the cellularity of the organ decreases with age more in men than in women, and there is a greater increase in glia formation in men than in women. The incidence of tumours in the organ is greater in boys than in girls, and these delay puberty when composed of parenchymal cells, but precipitate it when formed of other cells. The role of the organ in transmitting optic stimulation to the hypothalamus and through it to the pituitary has been described above. While pinealectomy prevents testicular atrophy in hamsters kept in the dark, it is not known whether gonadectomy affects the activity or structure of the pineal gland, nor whether the level of melatonin secretion varies with sex.

The weight of the thyroid in proportion to the body of males and females appears to be the same, but this does not preclude a sex difference in the ratio of stored colloid to active follicular tissue. The latter is suggested by functional differences, which are usually measured by the uptake of labelled iodine. This is greater in female rats than in males, varies with the oestrous and menstrual cycles and during pregnancy and lactation, and is reduced by ovariectomy. Female mice release a greater proportion of the thyroid hormone into the circulation than males, and have thus a lower ratio of the ^{131}I count for the gland to that in the serum. The size and activity of the gland depend on the intake of iodine in the water and food, which, if it is insufficient, is one of the causes of endemic goitre. Goitres can also be caused by excessive intake of goitrogens (substances which interfere with the metabolism of iodine and the synthesis of thyroid hormone), which are present in swedes, kale, cabbage and turnips, as can be demonstrated in rabbits and rats.

Mild and severe abnormalities in thyroid function are more prevalent in women than in men, and enlargement of the thyroid is noticed in 1% of the male and in 9% of the female population in the North of England. Primary hypothyroidism has a sex ratio of 1 man to 7 women, and in 80% of the patients autoimmune antibodies to the thyroid can be found. The serum level of TSH is twice as high in women of reproductive age as in men or in postmenopausal persons. In rats, however, the TSH level is higher in the male than in the female. The uptake of labelled iodine compounds varies with the oestrous cycle, and is superimposed on the circadian rhythm. Oestrogens increase the rate of thyroid secretion in male, female and gonadectomized rats. Prolonged treatment with oestrogens of rabbits depresses the thyroid function. The effect of thyroxin on the metabolic degradation of hexobarbitol by liver microsomes of the rat varies with sex.

Thyroid deficiency in children inhibits the maturation of the brain and leads to cretinism; in rats it slows down the increase in number of brain cells during the first three postnatal weeks, but prolongs mitotic activity beyond this period. It also interferes with the differentiation of the brain and alters its wiring pattern. Hyperthyroidism increases the metabolic rate, the gastrointestinal activity and motility, the rate of respiration and of the heart beat, and the stroke volume. It induces a loss of weight, but increases the appetite and thirst, causes an intolerance of heat, excessive sweating and flushing, muscular tremors, irritability and emotional disturbances. All thyroid disorders are more prevalent in women than in men.

In addition to the iodinated hormones, the thyroid produces in its 'C-cells' calcitonin, which with Vitamin D and the parathyroid regulates the absorption, deposition, release and excretion of calcium in the intestines, bones and kidney. The production of calcitonin is not influenced by TSH, but responds to the plasma level of Ca^{2+}. Administration of oestrogens lowers the sensitivity to calcitonin, but it is not clear whether this effect is due to an action on the C-cells or the peripheral receptors.

The parathyroids produce a hormone, parathormone, which releases calcium from the bone, while calcitonin inhibits this release. Vitamin D facilitates the binding of calcium, and together with the two antagonistic hormones regulates the calcium and the linked phosphate metabolism of the body. These three factors are responsible for both the excessive calcification and ectopic ossification and the alternative depletion of the bone salts in osteoporosis, which leads to an increased fragility of the bones, particularly in elderly women. This suggests a sex difference in the metabolism of the bones, but the specific agent has not yet been identified.

Whether the thymus of mammals has endocrine functions outside the immunological field is not clearly established. Thymectomy of mice in the neonatal period reduces their growth and 'nude' mice tend to be stunted, but these effects may be the sequence of immunological incompetence. The thymus has been credited for a long time with a growth factor promoting the increase in size of tadpoles and inhibiting their maturation. This effect is opposed to that of the thyroid, which promotes differentiation, reduces proliferation and accelerates metamorphosis. This effect of the thyroid on the development of mammals is substantiated, while the opposite effect of the thymus is not.

The islets of the pancreas contain three distinct types of cells with endocrine functions: the B-cells secreting insulin, the A-cells glucagon and the D-cells gastrin. Outstanding among its many functions, insulin lowers the level of blood sugar, while glucagon raises it. Gastrin stimulates the secretion of acid in the stomach and that of insulin in the pancreas, while glucagon inhibits the acid secretion in the stomach. There are some indications for sex differences in endocrine functions. After treatment of male rats with progesterone for three weeks, isolated islets produce more insulin than controls. Progesterone increases the mass of B-cells and insulin output in male and female monkeys; in bitches and in spayed rats the same result is obtained by testosterone. Oral contraceptives containing progesterone enhance the response to insulin in women, while oestrogens lower the level in acromegaly. Diabetes mellitus linked with obesity is prevalent in middle-aged women. Men have a higher level of blood glucose than women, but this is due in addition to insulin to glucagon, growth hormone and corticosteroids, and varies during the day with the intake of food. Gonadectomy has no consistent effect on the pancreas.

The sex differences of the adrenal glands are well documented. In most mammals the female adrenal is proportionately heavier than the male organ, and this is due largely to the cortex, which accounts for 90% of the weight in man, rat, cat and mouse. The hamster is an exception; the male gland is relatively larger than the female organ and is composed of medulla and cortex in equal proportions. The mammalian foetus has an adrenal larger in proportion to the body than does the adult. The large foetal cortex involutes in the neonatal period, and is replaced by the definitive cortex, which develops and remains under the influence of the pituitary. Distinct from the foetal cortex, the murine adrenal has a juxta-medullary X-zone which involutes in males at puberty and in females during the first pregnancy. Castration of mice, voles, hamsters and cats induces the formation of an X-zone, and in rats an enlargement of the cortex without stratification.

The production of glucocorticoids in the inner zones of the cortex is controlled by ACTH, while aldosterone is synthesized in the outer regions of the zona glomerulosa, and is regulated by the renin-angiotensin system. The renin is produced in the juxta-glomerular zone of the kidney, acts on globulins in the circulation and helps to form angiotensin. There is no evidence for sex differences in the secretion of aldosterone. In addition to cortisol and aldosterone the cortex produces testosterone, small amounts of oestrogens, and androstenedione, which can be converted into oestrogens outside the adrenal. Men produce 20 times as much androgens as women, while the female secretion of androstenedione is 2·5 times that of males.

Oestrogens stimulate the production of ACTH in the pituitary, and so lead to an increase in the size and secretory activity of the cortex, while ovariectomy induces shrinkage of the adrenal and reduced activity, and thus resembles the effect of testosterone. Castration increases the size and activity of the adrenal. Oestrogens also act directly on the gland. The circadian rhythm in adrenal activity is related to that of ACTH, but persists in the isolated gland kept in organ culture. The rate of inactivation of glucocorticoids in the female liver is 3 to 10 times that of the male. In many species of wild rodents the adrenals enlarge at the start of the breeding season, though in some this increase is restricted to the females, while the male organ shrinks. These changes may be related to the increases in the production of oestrogens and androgens respectively. The production of androgens is responsible for the virilism and hirsutism of menopausal women, and the absence of such effects in diseases of the adrenal cortex implicates the adrenal as site of androgen production.

The adrenal medulla synthesizes the catecholamines epinephrine and nor-epinephrine. The latter is produced also in the brain, in sympathetic nerve endings and chromaffine tissue. The proportion of adrenal epinephrine to nor-epinephrine is 10 to 1. There is no evidence for sex differences in the activity of the adrenal medulla and the removal of this part of the organ affects mainly the carbohydrate metabolism.

The endocrine functions of the gonads and their effects on the other constituents of the body are some of the major subjects of this review. Many facets of their influence have been discussed under various headings and need not be repeated here. It should be stressed, however, that in both sexes the gonads build their specific hormones from the same precursors, and that the testis can synthesize oestrogens and progesterones and the ovary androgens. Indeed the formation of testosterone by the theca cells of the ovarian follicle is necessary for the synthesis of oestrogens by the granulosa cells, at least in the sheep. The amounts of androgens produced by the ovary, and those of oestrogens and progesterones by the testis, are small compared with that of their own specific secretions, and their effectiveness is reduced by their being bound to proteins and by the lack of receptors in the target organs. Adrenal and ovarian androgens are thought to be necessary for the development and maintenance of the sexual drive in females, but the amount of androgens in the plasma and urine of women is only a tenth of that in men.

The effects of gonadectomy and the similar results of androgenization of females and feminization of males by hormones or antihormones vary with the stage of development and decrease from the foetal to the neonatal and later periods. The external genitalia and secondary sex characteristics remain sensitive to hormonal manipulation even in adult life. This statement applies to those features which require hormonal stimulation or their absence for their maintenance, as distinct from the induction of sexual characteristics by chromosomal or hormonal action in the earlier stages. The development and shrinkage of the mammary glands, the sexually dimorphic hair distribution and alopecia, are examples of the persisting hormonal control, and so is the sexual drive. The chromosomal abnormalities resulting from XO, XXY, XYY constitutions and their sequelae vary with species, and their detailed description is beyond the scope of this treatise.

The production of hormones appears to be widespread throughout the body, and new active compounds are continuously added to the inventory of such substances. Even the major hormones may be produced by unusual tissues, as is manifested in the ectopic endocrine secretion of tumours derived from normal parent tissues that are devoid of such activity. Thus ACTH is found in tumours of the lung, thymus, pancreas, and prostate, and more frequently in male than in female patients. An insulin-like factor has been observed in various abdominal sarcomas and hepatic tumours, while a parathormone-like substance may be secreted by cancers of the lung, kidney, ovary, uterus and pancreas, and TSH by gastro-intestinal tumours, choriocarcinoma and testicular teratomas, and vasopressin (ADH) by bronchogenic carcinomas. Precocious puberty may be induced by hepatoblastomas of boys secreting chorionic gonadotropins, and a hyperplasia of the breast in males by the secretion of gonadotropins by lung cancers. Similarly tumours may have receptors for hormones which are not present in the normal parent tissues. It is assumed that during the malignant transformation of the cells a histone repressor is inactivated, or a regulator gene deleted, thus allowing the manifestation of an inherent, but normally repressed, capacity for the production of the endocrine factors, and alternatively for that of receptor sites.

13. The central nervous system and the sense organs

Human, equine and murine brains are proportionately larger in the female than the male (tables 6 to 8), and the same may be true for other mammals. The maturation of different brain regions varies in time and developmental stage, and is earlier in girls than in boys. Thyroid hormones, corticosteroids, androgens, oestrogens and nutrition affect the growth and the maturation of different parts of the brain of rats. The effect of these agents is most evident in the development of the sexual dimorphism of the hypothalamic centres. Since the hypothalamus is closely associated with other parts of the central nervous system, some of the sex differences are bound to be reflected in them, and some have been found. It is probably more unjustified than for any other organ to extrapolate observations on the brain from one species to another, especially to the more advanced human brain. The direct application of surgical, biochemical, electrophysiological, electron microscopic, pharmacological and radioimmunological assays under defined experimental conditions is not possible in man, though such investigations in animals, and particularly in rats, cats and monkeys, as well as dogs, have adduced a flood of new facts and concepts in recent times. Some of the insights thus gained into the fundamental processes of brain development and functions are of significance for the understanding of the human brain, and provide a framework for the interpretation of psychological tests and for clinical, neuropharmacological and electroencephalographic studies in man.

Details of male and female psychology and sexual behaviour are outside the scope of this essay, and are amply covered by experimental psychologists. A few remarks about psychosomatic differences with sex seem appropriate. Boys are usually better at visuospatial skills, and girls at verbal formulations. This appears to be due to a preferential specialization of either the left or the right half of the brain; the left controls speech and verbal skills, the right the visuospatial performances. The left half appears to mature earlier in girls than in boys. Whether the time of maturation affects the intelligence of the two sexes is a controversial subject. The two halves of the brain are linked structurally and functionally, and to some extent loss of function in one region can be compensated by other parts.

Aggressive behaviour of males is related to androgen levels, as becomes obvious in the combativeness in the breeding season, and applies also to species capable of mating irrespective of the season. Dominant females appear to have higher levels of progesterones than the subordinate members of their sex. That an XYY constitution of men has any bearing on aggressiveness and criminal tendencies remains an unproven conjecture. The male cat, rat, rabbit and guinea-pig need an intact cortex for the performance of the reproductive function, while the female does not. On the other hand, an olfactory bulb is required by female mice, but its removal does not interfere with the mating process of males. Psychological, intellectual and socio-economic factors affect human reproductive behaviour more than that of other mammals.

Clinico-pathological investigations have localized specific functions in circumscribed regions of the cortex, for instance the control of movements, the perception of sensory stimuli, the analysis of optical and olfactory perceptions. Metabolic differences in enzyme activities, in rate of conversion of impulses, and in concentrations of neurotransmitters, and their variation in lesions and abnormalities in various regions have been recognized within the last 30 years, but in most instances little attention has been paid to sex differences. New substances produced in the brain, such as enkephalin and endomorphine, have been discovered fairly recently, and so has their association with the control of the secretion of growth hormone and prolactin. Whether they are sexually dimorphic and play a role in disease processes such as schizophrenia is not yet known. Specific blocking agents are used to inhibit the localized action of dopamine and serotonin, and thus to investigate their role in physiological and pathological conditions. Sexual dimorphism is suggested by the prevalence in men of lesions of the corpus striatum, causing Parkinson's disease, which is due to a deficiency of dopamine and manifests itself in muscular rigidity and tremors. In such cases, dopamine is used as a therapeutic agent. Research on such lines is certain to reveal more sex differences in the structure and function of the human brain.

The sexual dimorphism of the neuroendocrine centres which control the pituitary and general endocrine activity of the body is well documented. These regions control the female cycles, puberty and the onset of the breeding period, as well as the conditions of pregnancy, birth and lactation. Many differences in cytology and enzyme levels beween males and females are known, and in the latter there are variations with stage of the cycle, pregnancy and lactation. In the rhesus monkey the lining epithelium of the hypothalamus has a double layer of cells separated by a space lined with microvilli only in the mature male, while in the female it has bulbous projections into the third ventricle which vary in length and shape with the menstrual cycle. The preoptic area of female rodents and monkeys has larger nuclei, a higher incidence of non-strial synapses on the dendritic spines of the neuropil, a greater activity of oxidative enzymes and more 5-HT. Metabolic variations in this region, as for instance in the uptake of labelled methionine have been mentioned as varying with the circadian and oestrous rhythm (chapter 10). Almost all these features can be reversed to the male pattern by androgenization of females during the critical neonatal period in rodents. They are also affected in castrated or spayed animals.

In rats labelled oestrogens are bound by receptors in the nuclear fraction of the hypothalamus of females, but not of males. These receptors develop in the female during the third and fourth postnatal week, and thus reduce the sensitivity to androgenizing procedures, and later terminate the critical period for fixing the sexual dimorphism of the hypothalamus. The examples of structural and functional differences with sex of the centres in the brain stem and hypothalamus can be multiplied, but a detailed account of them must be left to the special literature.

The control that the neuroendocrine centres exert on the pituitary activity and through it on the gonadal and general endocrine system and on the female cycles, is best shown in transplantation experiments. If male or female pituitaries are grafted into the brain of male or female compatible hosts, the sex of the host determines whether the transplanted pituitary secretes gonadotropins according to the cyclical female or the uniform male pattern. If the male or female pituitaries are grafted together with their own hypothalamus, the sex of the donor determines the pattern of secretion of gonadotropins in both male and female hosts. The control of the secretion of gonadotropins is attributed to a dual function of the hypothalamus; a tonic secretion is regulated by cells in the arcuate nucleus and the ventromedial region, and this is alike in males and females. A cyclical discharge in females is controlled in the preoptic region of the hypothalamus. This concept assumes a sexual dimorphism in the central control of gonadotropins, which is evidently more complex and stricter in the female than in the male. This is confirmed by observations on pituitaries placed in ectopic positions; only those derived from females produce excessive amounts of LH, while male-derived ones do not. Of the gonadotropins FSH and LH are separately controlled centrally and by feed-back in the female only, where they have distinct functions in a temporal sequence, while in the male FSH and LH have synergistic functions and lack separate regulation. In hypophysectomized males the testicular function can be maintained by LH alone.

Androgens implanted into the region of the arcuate nucleus affect females more than males and, varying with dose, inhibit ovarian secretion or cause permanent damage to the ovary. In general, oestrogens are more effective than androgens in affecting the production and release of LH. The feed-back from the periphery to the neuroendocrine centres and the pituitary is essential for the maintenance of the male and female pattern of activity and in the latter for the sequences of the processes involved. These and their superimposition on the circadian rhythms have been discussed in outline (chapter 10). They are also dependent on the connection of the hypothalamus with other parts of the brain and with the neurotransmitter system. The latter contains steroid- and gonadotropin-sensitive elements which are involved in the signals for the discharge of the releasing and inhibiting factors produced in the cells of the hypothalamic complex.

A combination of delicate electrophysiological techniques with immunofluorescence, electron microscopy and histochemistry has made it possible to recognize some cells in various regions of the hypothalamus that regulate the production and release of the substances governing pituitary activity, and to trace back their connections with other parts of the brain. A number of detailed sex differences have been revealed in these investigations, as might be expected from the differences in the fundamental patterns of activity. Though the individual cells have specialized and discrete functions in the control of gonadotropins and other pituitary hormones, their interconnections are responsible for some spread of the stimulating impulses. Thus gonadotropins, as well as ACTH, TSH, growth hormone and prolactin are released at some phases of the oestrous and menstrual cycle, during pregnancy or lactation or after administration of oestrogens. The superimposition of such surges on the circadian rhythms accounts for some of the sex differences due to endocrine activity in the genital organs, in the secondary sex organs such as the breasts, in the skin glands, thyroid, lungs and thymus, and in temperature and rate of metabolism. The effects of the photoperiod, ambient temperature, and presence of males or females on the periodic events of oestrus and menses, and of the breeding seasons, as well as on the onset of puberty, have been dealt with.

The oestrous and menstrual cycles affect the attitude of the females to males, and this is true even for women as shown in the premenstrual tension. A variety of optic, auditory and olfactory stimuli may be disregarded by either sex outside the breeding season, or by females not on heat, but serve as means for sexual arousal in a hormonally receptive female. The effectiveness of sensory stimulation by sight, smell or sound varies with species. In man optic stimuli, intellectual qualities, changing fashions in the concepts of beauty, and socio-economic aspects are some of the factors operative in the choice of desirable mates, while animals have to rely on their individual endowments of size, strength, agility and special adornments such as antlers and manes, while many species rely predominantly on olfactory stimuli. Such pheromones are produced by rams and billy goats, and stimulate ovulation in females that have been without male companions. The introduction of a male into a cage of female mice synchronizes their oestrous cycles. Cats show oestrous behaviour when put into a container previously occupied by a tom. The smell of the boar is necessary for the successful mating of pigs, and removal of the olfactory bulb of female mice, but not of males, impairs their reproductive performance. The male pheromones are not produced in castrates, but can be elicited in spayed females by treatment with androgens. Female pheromones arouse the male, and rams can be excited sexually by swabbing fluid from animals on heat on pregnant ewes. The smell of urine from strange males incites aggressive behaviour in male mice, and blocks the pregnancy in females by delaying implantation. This block can be prevented by injections of prolactin or by anaesthetizing the hypothalamic centres with reserpine. Mature women are more sensitive than men or girls before puberty to some smells such as synthetic musk or some urinary steroids. Ovariectomy reduces this sensitivity, which can be restored by administration of oestrogens.

Female rhesus monkeys are most sensitive to odorants in the preovulatory period and this receptivity is correlated with the plasma level of oestrogens. The olfactory epithelium at this stage forms olfactory knobs which are five times as long as those in males or in menstruating females. These changes in the ultrastructure of the olfactory epithelium vary with the stage of the menstrual cycle, respond to hormonal stimulation, and are suppressed by ablation of the olfactory bulb.

The sense organs are extensions of the brain, and sex differences in the perception of optic and auditory stimuli may be due to differences in the eyes or ears or to the analysis of the transmitted messages in the cerebral centres. The differences in visuospatial, largely optical, abilities, and in verbal performances, largely auditory phenomena, in boys and girls have been mentioned. Dyslexia, a disability to read due to a central abnormality, is more frequent in boys than in girls. Women are supposed to have sharper hearing than men, but are also more prone to deafness from otosclerosis, an ankylosis and calcification of the ossicles of the middle ear. The tolerance of bright lights and the ability to see objects in dim lights or at longer wavelength (red light) is reported to be better in women, while men are credited with the ability to discern minute differences. Male rodents are more susceptible than females to audiogenic seizures.

The sex differences in inherited and congenital malformations of the brain and the sense organs have been discussed in part I.

14. The skin and its appendages

The skin of mammals is sexually dimorphic in the structure and function of its components: the thickness of the epidermis, dermis and subcutis, the type, distribution and pigmentation of the hair coat, the colour of the skin, and the size and activity of the sweat, sebaceous, and the specialized apocrine and mammary glands. These sex differences vary with species.

The epidermis is thin where the hair coat is dense and the fibres are fine, but thick in hairless regions such as the palms and soles, where the layers of keratin are numerous. The type of the hair coat varies with sex, and as it is usually denser and finer in females, the epidermis is thinner and has less keratin than in males. The male dermis is made up of coarser and more numerous bundles of collagen fibres, but the subcutaneous fat layer is thinner than in females. The deposition of lipid is increased in spayed or castrated mammals, and the site of preferential accumulation of fat varies with sex. The sex differences become manifest at puberty, and increase with age and in the menopausal period. The fingerprints, characteristic for each individual and due to the ridges on palms, soles and digits, are broader and have more whorls and fewer arches in males than in females.

In male rats the epidermal thickness is controlled by the antagonistic action of testosterone and thyroxin; in thyroidectomized castrates testosterone increases the number of epidermal cell divisions and the thickness of the skin, while thyroxin inhibits mitosis and decreases the width of the epidermis. The effect on cell production may be supplemented by changes in the life span of the cells. Sex differences in the life span of epidermal cells are likely, since the epidermal and dermal changes are coordinated with those in the hair cycle, which have been established as differing with sex.

The colour of the skin is due to the presence of pigment cells, to the amount of vascularization, and to the thickness of the keratin layer and of the hair coat. In the human skin the number of melanocytes per unit area is approximately equal in males and females, but the amount and distribution of pigment granules within them varies with the exposure to light, and is governed by the melanocyte stimulating hormone (MSH) of the pituitary. Oestrogens increase the secretion of MSH, and the hormonal changes during pregnancy lead to the increased pigmentation of the areola of the breasts, and to the appearance of freckles in the face. The sexual skin of female baboons, rhesus monkeys and chimpanzees varies in colour and blood supply with the cycle.

Though the number of hair follicles in comparable areas is alike, their size, and the hairs produced by them, vary with sex in type, length, width and colour. The foetal lanugo hair is replaced by the fine vellus of children and women, and the latter in specific regions by the coarser terminal hairs. The regional distribution of the hair types diverges at puberty; terminal hairs appear in the face and on the chest of men, while axillary and pubic hairs develop in both sexes. The area of pubic hairs on the abdomen of women ends in a straight transverse line, but in men it continues towards the navel in the midline. Scalp hairs are replaced by vellus hairs in alopecia, but the number of follicles remains constant though some of them may atrophy. The length of hairs varies with the rate of growth and the duration of the growth phase (anagen) in the hair cycle, and both vary with sex and endocrine conditions. After a rest period the hair is shed, and a new cycle is started. The cycle of growth and moult of follicles may proceed independently of that of neighbouring hairs as in man, or may be synchronous as in many rodents, whereas in mice and rats the wave of hair cycles moves from the ventral region towards the dorsum and the head. The duration of the cycles varies from about three weeks in mice to five or six weeks in rats. The life span of hairs varies with their length from days, as in mice, to years for the scalp hairs of women. The hair cycle can be initiated by plucking of hairs, which starts a regenerative activity in the follicle, but cutting of hairs does not. The cycles in cellular activity of the epidermis and follicles are like those depicted in fig. 6.

The variation of the hair cycles with the endocrine conditions is evidenced by the spring moult in seasonal breeders such as hares, which coincides with the resumption of the sexual activity. A pregnancy and the period of lactation prolong the anagen in women, mice and rats, but shorten it in ferrets, and reduce the rate of growth of wool in some breeds of sheep. The endocrine system controls the hair and the sexual cycle, and there is some overlap, but a considerable difference in the duration of the cycles.

Testosterone promotes the increase in length and thickness of hairs in rats. After puberty the hair coat of male rats is longer and coarser than in females and less dense. In spayed rats the hairs grow in length, but this growth is inhibited by administration of thyroxin. The metabolism of the hair follicles varies with the region of the skin, and is controlled by the level of sex hormones. Male *Macaca speciosa* are bald in the head region and the follicles in these regions have a high rate of testosterone metabolism. Baldness is induced selectively in the same areas of females by prolonged treatment with testosterone.

Many of the sexual differences in the hair coat are localized to the head region: the manes of male lions, the dorsal crest of boars, the beards of men and baboons. After the menopause the hirsutism of women is largely restricted to the face, and to the regions covered by terminal hair in men. Adrenal insufficiency prevents this hirsutism, and before the menopause inhibits the development of axillary and pubic hair. Testosterone given to boys before puberty, and to female sheep and rats, coarsens the hair coat. Castration as well as treatment with oestrogen results in a shorter and finer hair coat in males. The rest-period of the cycle is shortened by gonadectomy, adrenalectomy, hypophysectomy and administration of thyroxin, and is prolonged by goitrogens such as methylthiouracil, by oestradiol and by corticosteroids. These effects are most conspicuous in those mammals which have synchronous hair cycles, such as many rodents. The regional variations in the response to testosterone levels are obvious in the replacement of vellus hairs in the face and on the chest of men, and that of terminal hairs on the scalp by vellus. There is also a marked difference in the timing of these contradictory actions, suggesting local variations in sensitivity at puberty and in early middle age.

The synchrony of the hair cycles and waves tends to become disorganized with age, but at different times in male and female rodents. The waves progress more slowly and are smaller with each generation of hairs, and cease in male mice at the fifth generation and in females at the sixth. Where the cycles are synchronous, the associated changes in the skin are easily recognized: the thickness of the epidermis increases at the start of the growth period, and that of the dermis and the adipose layer slightly later, while at the end of the period and during the resting phase (telogen) hair follicles, adipose layer, dermis and epidermis shrink. These changes are accompanied by those in blood supply and in size and activity of the attached sebaceous glands.

Colour genes carried on the X-chromosome are responsible for sex variations in the colour of hairs or of segment of hairs in females by the randomized heteropycnosis of either the paternal or the maternal X-chromosome, while the male has a more uniform pigmentation of hairs. In some hybrid strains of mice all hairs of the males are dark, but those of the females may be dark or light in different patches or may have light and dark segments in individual fibres. These chromosomal effects on the hair colour cannot be influenced by the application of sex hormones, or by the light or temperature variations that are responsible for colour changes in seasonal breeders. An example is the absence of pigmentation from the winter coat of arctic mammals, and the presence of pigment, or rather the spread of the granules within the cells, in the summer coat. The colour of hairs in even small regions responds to localized cooling at a critical time of the moult. The seasonal moults are conditioned by the photoperiod and temperature, which are perceived by the peripheral sense organs and relayed to the hypothalamus and the pineal body. These centres control through the pituitary the activity of the gonads, the adrenals and the thyroid.

The eccrine sweat glands vary in number with sex in the X-linked hypohidrotic ectodermal dysplasia; they are absent from men and reduced in total number in female siblings, being normal in regions where the normal X-chromosome is active and absent if it is heteropycnotic. Though no other structural sex differences are recorded for human eccrine glands, they differ functionally in the thermoregulation. Men sweat more profusely than women, who depend more on the alteration in blood supply to the skin and on respiratory activity to dissipate heat.

The male apocrine glands of the axilla, anogenital and mammary regions, and the more widely distributed glands of many species, develop during puberty, are controlled by sex hormones and are altered in size and function by gonadectomy. In women the width of the lumen and the height and activity of the epithelial cells of the axillary glands vary with the menstrual cycle, and reach a peak at ovulation and a trough at menstruation. In wild rabbits the apocrine glands of the submandibular, inguinal and anal region are larger in the male than in the female, are reduced by castration or oestrogen treatment, and are restored in castrates by testosterone application. Dominant bucks have particularly large glands.

The sebaceous glands are often, but not always, associated with the hair follicles. They are active in the human foetus between the 13th and 15th week, regress in infancy, and become active at puberty. In immature boys they respond to androgenic treatment. In most mammals they are larger in males than in females, and are stimulated by testosterone and inhibited by oestrogens. The secretory activity is decreased by oestrogens, but antiandrogens additionally suppress the mitosis necessary for the replacement of the holocrine cells. Before puberty the glands are not sexually dimorphic, and topical application of testosterone at this phase to rats, rabbits, hamsters and mice induces mitotic activity, and hastens the turnover of cells and secretion. Women produce more sebum during the progesterone phase of the menstrual cycle, and large doses of progesterone increase the secretion in female and castrate rats, probably because of the androgenic properties of the hormone. Testosterone and progesterone enhance the sterol metabolism in the skin of mice, and so does oestradiol, though it reduces the secretory activity of sebaceous glands in males. Cortisone has an effect comparable to that of oestrogens. Corticosteroids such as prednisone do not affect the secretion of sebum in men, but reduce it in women by 20% and in castrates by 40%.

The structure of sebaceous glands varies in complexity with species and region and some specialized forms occur on the ventral pad of gerbils, as brachial glands of lemurs, and as flank organs in hamsters. They are larger in males than in females, and respond more strongly than simple glands to testosterone stimulation. The side glands of the shrew are oestrogen-dependent as regards size and secretion. All the sebaceous and apocrine glands produce pheromones to attract the other sex and to warn off rivals.

The mammary glands are specialized skin glands, and their embryonic development owes more to the absence of an inhibitory effect of androgens than to promotion by the female sex hormones. They enlarge at puberty in females and normally become functional only in them under the influence of prolactin, oxytocin and the female gonadal hormones. Oestrogens cause the enlargement of the mammary glands of men when given therapeutically or absorbed accidentally during the manufacture of the substance. The male rat has no nipples, but their development can be induced even in the adult by the administration of antiandrogens. Male rats and male guinea-pigs respond to treatment with oestrogens in the presence of prolactin, growth hormone and ACTH, with enlargement of the breasts and lactation. Milk production is maintained by spayed milch cows, i.e. in the absence of ovarian hormones.

The antlers of stags are conspicuous examples of sexual dimorphism of an appendage of the skin. They are formed at the start of the breeding season and are shed at the end and are dependent on the level of testosterone.

15. The respiratory and digestive organs

In addition to its main function of gas exchange, and in particular of carbon dioxide for oxygen, the respiratory tract is involved in thermoregulation, the production of sounds, and in facilitating olfactory stimulation. Sex differences in the mechanism of thermoregulation have been mentioned, and those in the pitch of voice are well known. Sniffing involves the respiratory activity in bringing volatile compounds into an increased contact with the olfactory epithelium, and some sex differences in this respect have been described. The functional role of respiration varies with species, and with the adaptation to environmental conditions, in particular the climate, the oxygen tension, which decreases with increasing altitude above sea level, and with the preferred mode of intraspecific communication, whether optical, auditory or olfactory.

The distinctive pitch of the voice of males is produced by the enlargement of the larynx under the influence of testosterone at puberty, or at the onset of the breeding season; the thyroid cartilage grows, alters the configuration of the larynx, and by lengthening the vocal cords lowers the notes produced. In the absence of this enlargement the voice of women and castrates retains a higher pitch.

The tracheal epithelium of women appears to change with the level of hormone present at different phases of the menstrual cycle. With higher oestrogen levels the nuclei occupy the middle regions of the cells, but migrate towards the apex as the progesterone level rises. This migration may be coordinated with the rate of mucus production and secretion in the trachea.

The lungs are proportionately larger in men than in women, in mares than in stallions, and about equal in male and female mice (table 6). The proportion of the lungs to body weight of males and females changes with developmental stage in man (table 7) and mice (table 8). In both species the weight is relatively greater at first in females, but at later stages the male organs grow faster than the female lungs in proportion to the whole body. The same pattern obtains even more markedly in the postnatal development of the heart.

The total capacity of the lungs in men exceeds that in women, and the average sex ratio is 1·41; similarly that for the vital capacity, which is the volume of air measured between the deepest inhalation and exhalation, amounts to 1·52, and for the breathing capacity to 1·35, while the sex ratio for the height in the group tested is 1·08 and that for the weight 1·27.

The rate and manner of breathing change at puberty; it is faster and shallower in women and of costal type, performed by the movement of the rib cage, while it is predominantly of the abdominal type in men, who use the diaphragm in addition to the rib cage. The muscles of the diaphragm in males are better developed than in females. This difference in respiratory movements may be related to the different mode of thermoregulation. Women sweat less profusely than men and rely more on panting for the dissipation of heat. The same connection is found in the costal type of respiration in dogs who rely on panting helped by the protrusion of the tongue, and the abdominal breathing of horses and ruminants who are able to sweat.

The composition of the exhaled gases reflects sex differences in the metabolism. To give one example, after equivalent doses of ethanol male mice exhale more acetaldehyde than females.

Though qualitative or quantitative differences with sex are not recorded either for the gross or for the fine structure of the mammalian lung, there appear to exist functional differences in the sensitivity to injurious agents. This is suggested by the prevalence in men of cancers of the lung, of chronic respiratory diseases and of emphysema. The sex ratio is high in all countries, though the total incidence of the disorders varies greatly. The exposure of men to greater risks than women cannot wholly explain this divergence. In babies under one year the sex ratio for deaths from respiratory disorders is very high, and subsequently decreases with age. In England and Wales the death rate for babies under one year from respiratory diseases has declined to one quarter between the years 1940 and 1967, but the sex ratio has remained at 1·31 in spite of the decrease in total death rate due to the improvements in social and environmental conditions and the introduction of new therapeutic agents. Such data suggest a greater susceptibility of the lung of boys to disease processes combined with a delay in the maturation of the immune defenses.

The sexual dimorphism of the teeth, the salivary glands, the liver and the gall bladder is well documented, but data about such differences in the oesophagus, stomach and the gut from the duodenum to the rectum are at best scanty. Since the prevalence of disorders in various regions of the human intestines varies with sex, it seems likely that subtle physiological variations with sex play a predisposing role in these processes.

The mammalian dentition consists of teeth with differing shapes and functions and is adapted to the needs of every species. Well known sex differences are the larger canines or incisors in, for instance, boars, male monkeys and elephants. Stallions have canines while mares do not, and the canines are enlarged in many male deer.

The submaxillary of rodents has attracted most attention amongst the salivary glands. It is twice or three times as large in the male mouse, rat and vole as in the female, and has more tubules in proportion to acini. The secretory tubules are sexually dimorphic, being much longer in the male than the female, with higher epithelium and greater secretory activity, which differs also in type. Only the male mouse produces an epidermal and a nerve growth factor and secretes also more proteolytic enzymes than the female. This special section is dependent on androgens and is reduced in length, height and activity in castrates. Female rodents respond to androgens with some elongation of the tubules and with greater secretory activity. The acinar mass consists of serous and mucous cells, the proportions of which vary with sex: male pigs, mice, hamsters and rats have more serous and fewer mucous cells than females. Some of the enzymes are produced only in the male, such as the hydrosteroid dehydrogenase, which is produced by boars only. Unlike the proteolytic enzymes, amylase and kallikrein are not androgen-dependent. The female hamster has a larger gland than the male or the spayed animal, produces more sialic acid, and responds to oestrogenic stimulation with increased secretion of mucus.

The other salivary glands (the parotid and sublingual) have no such distinct sexual dimorphism even in mice and rats.

The liver is proportionately very much larger in mares and female mice than in males, and slightly larger in women than in men. The relative size differences persist from birth to the adult stage suggesting a sustained (table 7) or increased rate of growth (table 8) in the female.

The rat has been and is being used extensively for studies on the liver in respect of metabolism, enzyme activity, effect of hormones, action of hepatotoxic drugs, and biochemical and structural composition. Conclusions based on such data, including those on sex differences, can be extrapolated to other species only with caution and qualification. Thus specific 5β-enzymes are found only in the liver of male rats, but in those of males and females of other species. The enzyme activities assessed in homogenates from male livers of rats are greater in various dehydrogenases and in 20-oxoreductase than those derived from females, and are increased by addition of testosterone and inhibited by oestrogens. The activity of α-reductase in the female liver is three to ten times as great as in the male and is depressed in prepubertal males by testosterone. Castration of rats causes shrinkage of the liver lobules, and a decrease in the amount of basophilic granulations of the hepatocytes and in the number of binucleate cells. Testosterone restores the size of the liver, the number of binucleate cells, and the basophilic granulations. Male livers contain more glycogen, but less fat and vitamin A, than female organs, and starvation reduces the fat content more in male than in female livers.

Oestrogens are inactivated in the liver and the metabolites are excreted in the bile, gall bladder and urine. Impairment of liver function results in hyperoestrogenism which is manifested in men by the atrophy of the testes and by gynaecomasty, and in women by disorders of menstruation or by postmenopausal bleeding. Oestrogens are lipotropic, i.e. they promote the synthesis of phospholipids and are responsible for the difference in the pattern of lipids in women of reproductive age compared with those in the menopause and with men. In spayed mice oestrogens restore the percentage of binucleate hepatocytes and they stimulate the phagocytic activity of Kupffer's cells.

Androgens enhance the phosphatase levels of the rat liver, protect males, castrates and females against the deposition of fat, and inhibit the citrate synthesis in homogenates of female livers.

Liver damage due to the deposition of fat is greater in male than female rats kept on a diet deficient in choline, while ethionine induces fatty livers in female, but not in male rats. Of the many hepatotoxic agents acting preferentially on one or the other sex in rats carbon tetrachloride, *Senecio* alkaloids and tannic acid affect males more than females, while hexabarbitone, ethanol and trichloroethylene have a greater effect on females. Testosterone promotes protein synthesis in the liver and is thus able to counteract the effect of ethionine in females, while the sensitivity of males is increased by oestrogens.

Liver tumours are induced preferentially in female mice by the azo dye butter yellow (*p*-dimethylaminoazobenzene) and aminoazotoluene, and in male rats by 2-acetylaminofluorene. Spontaneous hepatomas occur more frequently in male mice and men than in females.

The rat liver is capable of regeneration after partial hepatectomy, and cell divisions involved in this process are subject to a circadian rhythm which has a peak at 0900 to 1000h and a trough at night.

The gall bladder empties faster in boys than in girls, but after puberty it is faster in women than in men. Gall stones, cholecystitis and pancreatitis occur more frequently in women than men.

The sexual dimorphism of the endocrine functions of the pancreas have been discussed; the eccrine components do not appear to vary with sex.

Sex differences in the structure and function of the oesophagus have not been recorded, but shortening of the oesophagus by muscular contraction, which is associated with the appearance of a hiatus hernia, is more common in women than in men. The structure and function of the stomach and the gut and the secretion of endocrine and paracrine peptides are likely to show sex variations, since they, as well as the rest of the digestive apparatus, break down the food and transform it into body constituents by metabolic processes, many of which are known to be sexually dimorphic. It will thus not be surprising if variations in secretory, synthetic and general enzymatic activity in male and female mammals are discovered. The acid production in the stomach is known to be greater in men than in women. Many diseases show a sex prevalence; gastric ulcers are more frequent in women, while duodenal ulcers predominate in men. The incidence of stomach cancers is many times as high in men as in women in all countries, irrespective of the total incidence of this disease. Such findings suggest a sex variation as the physiological basis for the pathological processes, but remain to be revealed.

16. The urogenital tract

The role of hormones and of chromosomes in the determination of sex and in the formation of the gonads and other parts of the genital tract has been the subject of part I. A detailed comparison of the many species variations and modifications of the penis, the clitoris, the scrotum and the vulva is beyond the scope of this essay, and this section will deal only with aspects of the sexual dimorphism of the kidney and the urinary tract.

The kidney is proportionately larger in men and male dogs, cats, rats and mice and in mares than in the other sex (table 6). Castration reduces the volume of the kidney in dogs, cats, rats and mice, but not that in guinea-pigs and hamsters. Following unilateral nephrectomy in dogs and rats the compensatory hypertrophy of the other kidney is promoted by testosterone, and is retarded by thyroidectomy and hypophysectomy, but not by gonadectomy.

In mice, the male kidney is proportionately larger from birth onwards (table 8), while the human kidney is initially larger in girls but has a lower growth rate in relation to the body after puberty than in males, and thus resembles in developmental pattern that of the lung and heart. The proportion of tubules to glomeruli in the renal cortex has a high sex ratio. Male cats have larger kidneys, but only 9000 glomeruli as compared with 23 000 in the female, and the rat has similarly 27 000 glomeruli in the male, but 35 000 in the female cortex. The epithelium of Bowman's capsule surrounding the glomeruli is sexually dimorphic in mice; the parietal epithelium is composed of cuboidal cells in the male and of flat cells in the female. Androgens increase the height of these cells and also the total amount of tubules. Oestrogens cause the formation of cysts at the border of the medulla and cortex in rats and induce renal tumours in the male golden hamster, but not in the female.

The amount of filtrate produced per unit weight of kidney is approximately the same in all mammals, but the composition of the urine varies with species, strain and sex. Thus the urine of mice normally contains protein while human urine does not. Male mice excrete more solids and twice as much protein as females, who drink more water. The composition of the urine is a reflection of the enzyme levels in the tubular system. The activity of alkaline phosphatase in the mouse is promoted by androgens and inhibited by oestrogens or by castration. The level of glucuronidase of males is twice that of females and increased by testosterone. In rats the enzyme pattern of the male kidney is shifted to a female type by treatment with antiandrogens. At 25 years the urine of men contains about twice as much 17-ketosteroids as that of women, and in most species the male urine has more creatine and creatinine than the female one, and the level is decreased by castration and increased in females by androgens.

The urine of men contains 30% more calcium than that of women. Differences with sex in the metabolism of minerals are reflected in the composition of kidney stones, which in men are formed usually by calcium oxalate and in women by magnesium ammonium phosphate. The metabolism of calcium is regulated by parathormone, calcitonin and vitamin D, and by a feed-back relation of the kidney with the adrenal; renin is produced by the juxta-glomerular apparatus and converted into angiotensin which stimulates the production of aldosterone and the latter affects the renal function. There are no reports of sex differences in the renin levels, unlike the documented higher levels of erythropoietin in males, which also is produced by the juxta-glomerular apparatus. Angiotensin is, however, involved in the hypertension caused by the occlusion of renal arteries and this disorder affects predominantly men. The amount of erythropoietin in the urine averages 2·8 to 4·0 units per day in men, 1·0 in prepubertal boys, and 0·9 in women. Castration decreases the output of the agent and testosterone increases it. Erythropoietin acts with a lowered oxygen tension on the precursors of the red blood corpuscles in the bone marrow and stimulates the differentiation of the cells. About 90% of erythropoietin is formed in the kidney. In pregnancy its level and that of erythrocytes is raised.

The pheromones in the urine of mice and other species are androgen-dependent. Some of them are present in the contents of the bladder while others are produced by the accessory glands of the urogenital tract. The bladder of male mice is larger in proportion to body weight than that of females.

PART III
COMMON FEATURES IN THE SEX DETERMINATION AND SEXUAL DIMORPHISM OF MAMMALS

Common features

The account of sex differences and their variations in many species in previous chapters is incomplete, because data on many items are not available for both sexes, even in man and in the small laboratory animals that provide most of the information. A picture of the scope, underlying mechanisms and *raison d'etre* of sex determination and sexual dimorphism can be formed only by extrapolating to other species the little that is known, and it is necessary to remain conscious of the pitfalls of this procedure and of the need for qualifying almost all statements. A single example will illustrate this point; on average, males of many species (man, rat, ferret, cattle, lions) are larger than females, but in others (rabbits, golden hamster, many bats, spotted hyena) the reverse holds true. While the growth varies with sex in almost all mammals, it does not always do so in the same direction. Nevertheless it is essential to analyze the factors responsible for the differential rate and duration of growth, and from there to proceed to find the modifying conditions. This attitude conforms to the generally accepted tenet that all living organisms use the same building blocks, though in a different manner and plan of construction. An investigation of the general factors accounting for variations with sex in different species is necessary for understanding the physiology and pathology of all mammals, and as a basis for conservation, preventive measures and therapy. where necessary. It is easier to define the common features in sex determination than those in the dimorphism of adults, which are only apparently minor modifications compared with the differences between species; a male elephant is larger than the female, but has more in common with her than with a male rat that is larger than the female.

The sexual development of mammals is completed after puberty, which follows birth within a few weeks in some species and after a considerable number of years in others. The development is controlled at successive, but overlapping, levels; it is initiated on the chromosomal level at fertilization, continued on the gonadal and ultimately regulated on the neuroendocrine level. Apart from the functional differentiation of the reproductive system, all somatic organs and systems are encompassed, as well as the psychology, social structure and adaptation to or choice of the environment. The urge to propagate, to ensure the survival of the species and that of the offspring, is fundamental, and mammals have to sustain their young in the period of gestation and during an often prolonged infancy. Water tortoises bury their fertilized eggs in the sand, and after hatching the young have to find their own way to the sea or river and to fend for themselves. Thus the maternal functions of mammals are important features for their sexual differentiation, though it does not reach the degree of specialization of bees with the distinction of the queen from the workers and drones.

For sexual reproduction, mechanisms must be provided for the maturation of the germ cells, their release, and the synchronization of male and female activities. The eutherian mammals have in addition to create an internal environment suitable for the growth and maturation of the embryo and foetus, for parturition and subsequently for lactation. These requirements are met by the differentiation of the reproductive organs, of accessory structures and of special controls integrating the multitude of metabolic processes in different organs, which involve special refinements in the endocrine and neuroendocrine regulations which are not necessary in the males. In many species many more males are produced than are needed for the purpose of propagation and survival of the species. In fact, in many animal societies some males never have a chance of mating and may even be treated as outcasts.

The life span of the germ cells in a fertile state is short, and this limits the period during which mating can be successful. Hence arises the need for the synchronization of reproductive functions in males and females; and the female dictates the timing. In the simplest form, she stores the sperm in her genital tract until she releases ova for fertilization—a method essentially similar to the practice of artificial insemination, but requiring special adaptation. In this instance the mating is divorced in time from fertilization. In some marsupials, in red deer and in some hibernating animals, fertilized ova may remain in a dormant state for some months. The periods of sexual cycles, whether oestrous or menstrual, and within them the short period of ovulation, limit the time for mating, and with it for fertilization. Rabbits ovulate after copulation, but many other mammals accept the male only at oestrus or the midcycle of the menses, though women do not restrict themselves to this stage of the cycle. The variations in this respect, as well as the associated psychological and behavioural phenomena, are too varied, in view of the adaptation of the species to its environment, for this survey, which concentrates on universal features of sexual reproduction in mammals. The cyclical activities of the female extend from the neuroendocrine centres to the gonadal and other endocrine organs, which control the metabolism and are responsible for the functional gender differences which contrast with the uniform level of the activity in males. Circadian rhythms are common to both sexes, but the sexual periodicity is superimposed on the circadian in females.

Sex differences are limited, but significant modifications in the development of species and the common factors between all their members far outweigh the variations between males and females. The genetic information is distributed over a varying but large number of chromosomes, and the sex chromosomes carry only a fraction of it. The constancy in the characters of a species is assured by the equal number of genes handed on by spermatozoa and oocytes in their autosomes. Of the specialized sex chromosomes the Y-chromosome determines the sex by its presence in males and its absence from females, but carries few if any structural genes. It is obviously not essential for the life of a mammal, since all females do without it. The X-chromosome is essential for life, and carries numerous structural genes—almost 100 in humans—which are present in double dosage compared with those on autosomes. Hence the male needs only a single X-chromosome and the female inactivates one of the pair, while males, like females, need both members of a pair of autosomes. Since the total number of chromosomes varies with the species, the genetic information must be distributed differently over them, and the X-chromosomes do not carry identical genes in all mammals.

The male receives his X-chromosome from his mother, and hands it on unchanged to his daughters, since, being single, it cannot pair with its like and exchange genes during meiosis, and instead forms an XY body while the autosomes pair and exchange parts. In the female the X-chromosomes pair during meiosis like the autosomes, and their gene composition is reconstituted between its reformed maternal X-chromosome and the X-chromosome from her paternal grandmother. There is thus a generation gap between the maternal and paternal X-chromosome of the daughter, while the son always receives the reconstituted maternal X-chromosome. Since the somatic cells need only one X-chromosome because of the dosage compensation, one of them is inactivated and appears as a Barr body in most tissues and as a drumstick in leukocytes. The heteropycnosis of one X-chromosome occurs in stem cells in early embryonic development, and the descendants carry the same active and the same inactive

X-chromosome. In most eutherian mammals the paternal or maternal X-chromosome is made heteropycnotic at random, but in marsupials the paternal chromosome is always inactivated. Thus the somatic cells of female eutherian mammals form tissues which are a mosaic of cells with paternally or maternally derived X-chromosomes, while those of the male are uniform in this respect. The condition is reversed with respect to gonocytes; the oocytes are homogametic and have new X-chromosomes as well as new autosomes thanks to the pairing process in meiosis. The spermatozoa are heterogametic and carry either an unchanged maternal X-chromosome or a Y-chromosome of the same constitution as that of his male ancestors, since there is no pairing during meiosis. The autosomes pair and exchange genetic information as in the female. The Y-chromosome carries few if any structural genes.

These are the general principles in the differentiation of the sex chromosomes. Abnormalities occur during meiosis or in the mitotic divisions following fertilization, and lead to translocations between autosomes, between autosomes and the X-chromosomes, and to non-disjunction of chromosomes resulting in uneven chromosome numbers, XXX or XO or XXY or XYY, and these constitutions are reflected in impaired fertility or other disorders. Normally occurring exceptions have been noted in some species where, as in the wood lemming, fertile females may be either XX or XY. These animals must have some mechanism for overriding the sex determining effect of the Y chromosome. It is not clearly established whether the Y-chromosome directly initiates the differentiation of the male gonad or whether it activates genes on the X-chromosome or the autosomes and thus acts indirectly.

The random inactivation of one of the pair of X-chromosomes in female cells has great practical importance. If one of them is faulty or carries a mutant gene, it will be active in half of the tissue cells and inactive in the other half, which may be able to compensate for the defect in females, as for instance in haemophilia and colour blindness; these are manifest in all males carrying the faulty maternal X-chromosome, but those receiving the normal maternal X-chromosome will be free from the disorder. Similarly the female may receive a normal X-chromosome from the mother and a faulty paternal one, if the father is afflicted. After heteropycnosis it will be inactive, but if it is active its defect can be compensated by the activity of the normal X-chromosome in the other cells composing the tissue. Thus the disorder will not be manifest in females, who also have the ability either preferentially to inactivate the faulty X-chromosome or to eliminate the cells carrying it. This ability is seen distinctly in the enzymatic activities of cells of various tissues taken from the same woman; in fibroblasts only 50% of the cells produce the enzyme, while all haemopoietic cells do so. The mosaicism of female cells with regard to the X-chromosomes enables them to cope with defects which are manifest in males, and gives them more scope in producing some of the proteins and enzymes, and thus to counteract some of the effects of injurious agents.

Whether by direct or indirection action, the Y-chromosome initiates the differential growth of the medullary region of the gonadal anlage on the mesonephros, i.e. the multiplication of the interstitial cells and their secretory activity in producing male hormones. The immigration of the germ cells appears to follow this initial differentiation of the gonads, which is earlier in males than the development of the cortical region in females. Testicular secretions carried via vascular anastomoses between the placentas to a female twin of ruminants are responsible for the development of freemartins by the inhibition of ovarian development in a female of normal XX constitution. The Mullerian duct derivatives and the external genitalia of such animals are of normal female appearance. The testicular secretion responsible is considered to be factor X or medullarin, since the injection of androgens into pregnant cows masculinizes the external genitalia, but does not impede the development of the ovary. Thus testicular hormones, rather than the action of the Y-chromosome, cause the atrophy of the cortical region of the gonadal anlage.

Castration, spaying, or treatment with steroid sex hormones can alter and restore somatic sex differences. The effects of castration are readily apparent in the suppression of pubertal changes, and comparable results follow the administration of oestrogens, while spaying or treatment with androgens impedes the development of secondary sex organs in females, but does not alter the genetic sex of mammals. The effectiveness of the procedures varies with the stage of development, but never overrides the definition of the genetic sex at fertilization. The ovarian hormones, oestrogens and progesterones, are produced in quantity only shortly before puberty, i.e. at a later stage of maturation than the androgens. The androgens or their absence in critical stages of foetal or neonatal development are believed to be responsible for the male or female differentiation of the centres in the brain, which becomes manifest after puberty in the cyclical activities of females and the steady level of that in males. The circadian rhythms of both sexes are controlled by the same regions in the brain. The critical period is in the foetal stage of monkeys, which have a fairly long duration of gestation, and in the neonatal period of rats and mice where the pregnancy is short. The differentiation of the centres can be altered in the first two weeks of life of mice and rats by treatment with male or female sex hormones, auditory or optic stimuli, or lesions in specific parts of the brain, and will become manifest after puberty by the delay in its onset, the inhibition of full gonadal functions and in females by the suppression of the sexual cycles.

The activities of the endocrine components of the gonads are regulated by their feedback relations with the centres in the brain via the pituitary and its hormones. The neuroendocrine centres in the brain are connected with the cortex and the pineal body, which transmit sensory signals from the peripheral sense organs, and these are monitored and integrated with the internal feed-back to coordinate the functions of all body systems. The centres with their endocrine and nervous connections are responsible for the onset of puberty and so of the final stage of sexual development, and for the resumption of sexual activities at the onset of the breeding seasons. The degree of the central control of the gonadal functions is stricter in females than in males, as might be expected from their additional tasks of the synchronization of the mating processes, and of the requirements of pregnancy, lactation and nursing. In this context the term 'control' does not imply a one-way issue of orders, since the feed-back relationship ensures a two-directional flow of information and activation. The central control means that the peripheral activities are monitored and adjusted via the increased or decreased release of the pituitary hormones which in turn are switched on by releasing or inhibiting factors of the hypothalamus in response to the peripheral signals.

The sex differences in the gonadal functions follow different paths in their development. The testis produces androgens in foetal life, reduces the secretion after birth and resumes it before and at puberty. After their migration to the testis the germ cells multiply for a time, and are subsequently quiescent until puberty, when they proliferate, mature and form spermatozoa. The oogonia multiply greatly in foetal life and reach their peak in numbers then; they mature and reach the prophase of the first meiotic division in the perinatal period. They decrease subsequently in number, but remain in the same stage of development throughout life, or until the menopause, unless used in ovulation. The interstitial cells of the ovary are quiescent during foetal life and in infancy, and start their secretory activity at puberty. Puberty thus initiates the secretory activity of the ovary and spermatogenesis in the male, while the effect on the ova and on the testicular secretions is less dramatic. In both sexes the sensitivity of the gonadal tissues to gonadotropins increases.

The feedback relationship of the ovary to the centres diverges from that of the testis. The ovary produces sequentially oestrogens and progesterones in response to the pituitary FSH and LH respectively, while the testis maintains a steady level of androgens, apart from circadian variations and quiescence outside the breeding season; they are produced under the control of a gonadotropin similar to LH and usually referred to as such. FSH acts synergistically and at the same time as LH on the testis and, apart from initiating the proliferation of spermatogonia A, is not essential for the maintenance of testicular functions after hypophysectomy, though LH is. The hypothalamic centres have to be organized in the female to regulate separately and in sequence the release of FSH and LH, while no separate control is required in the male. This differentiation occurs during the critical period, but manifests itself only at or after puberty. The exact mechanisms responsible for the pubertal changes in the neuroendocrines are not understood, but the timing is influenced by nutrition and the preceding rate of growth, and by environmental factors, including the presence of members of the other sex, the photoperiod, temperature and auditory, optic and olfactory stimulations.

Spermatozoa as well as ova need factors from the other sex to render them fully fertile; the oocyte depends on the entry of a spermatozoon for the completion of the second meiotic division. The spermatozoon passes through its meiotic divisions and the process of spermiogenesis in the testis, but becomes motile in the epididymis and fully fertile after capacitation in the female genital tract, i.e. it requires factors produced in the female for reaching full fertilizing capacity.

The number of oocytes in women decreases at each ovulation from the initial stock of 500 to less than 100 at shortly before menopause. In rodents more ova are released at shorter intervals and their number too dwindles, whereas the pool of spermatozoa is continuously replenished, and amounts in man to a size sufficient to release 3×10^8 per ejaculate. The chances for a spermatozoon to effect fertilization are thus infinitely smaller than for an oocyte being fertilized. Thus a faulty oocyte presents a greater risk than a faulty spermatozoon for handing on genetic or chromosomal abnormalities on the autosomes. Maternal, but not paternal, age is linked with some congenital abnormalities such as Down's syndrome. In this respect X-chromosomes are somewhat different, as the female can compensate for defects on one of the pair, while the male has only one of them. Hence sex-linked disorders are found mainly in the male offspring.

The incidence of chromosome abnormalities (0·3%) in boys is about twice that in girls. Congenital abnormalities comprising malformations in early embryogenesis and deformations of structures in later foetal periods vary with sex in severity, incidence and preferential site. The genetic component of some is revealed by the increased frequency of the same disorder in siblings or relatives. Most are due to autosomal mutations, translocations or another deficiency on one of the pair. Those of the X-chromosome are manifest in males and in homozygotic females, while the heterozygotic female can compensate for the deficiency. The total incidence of prenatal abnormalities and congenital disorders is difficult to assess, and with it any variation with sex. Many of the minor faults are neglected, and a sex difference in the severity of a malformation may lead to an early abortion of either the male or the female, while the other sex survives parturition and is recorded as a congenital abnormality. The sex differences in the prevalence of deformations are due in part to variations with sex in the rates of growth and maturation, but are largely due to as yet unknown factors. It is significant that teratogens are equally effective in both sexes, while similar spontaneous errors in development show a prevalence in one or the other sex.

The problem of the primary sex ratio, i.e. whether X- and Y-bearing sperm are equal in their chance of fertilizing an ovum and thus producing a sex ratio of 1·0, is still unresolved. The sex ratio at birth is greater than 1·0 in many species, and since more abortuses are found to be male than female, a high primary sex ratio is assumed, but contested. The sex of an abortus can be accurately identified only in the later part of a pregnancy and the cytological evidence of sex, though definite, is available only for very small samples. At this stage of our knowledge no definitive conclusion is possible on this issue.

Androgens have an anabolic function and promote growth by the proliferation of cells, which to some extent conflicts with and impedes their differentiation. Thus males tend to be larger, but less mature than females at birth. The difference in the rate of growth may be maintained in the early post-natal period, or reversed, and the gap tends to decrease during infancy, but widens at puberty. The increments in height, length and weight fall steadily after birth, but may be sufficiently great in small animals, that become sexually mature early on, to obscure a spurt in growth associated with puberty. The human data indicate beyond doubt that the female puberty precedes that of males, possibly because girls grow more slowly *in utero*, but mature earlier than boys and continue to do so in infancy. Though it is difficult to date the onset of puberty accurately, particularly in boys, the complex somatic alterations take a long time to develop, and in boys as in girls of the same age appear first in the taller and heavier individuals. In animals environmental factors of nutrition, optic and olfactory stimulation, social conditions and presence of the other sex advance or delay the pubertal changes. The first oestrus and the first menses are easier to diagnose than the stage of spermiogenesis. While the changes in the reproductive organs at this period are of about equal duration in boys and girls, the activation of the more complex central regulatory activities in girls may require more time. During the last century the menarche has occurred at a decreasing age of girls. Whether the advance of pubertal changes in boys is similar is not well documented. It is likely, since the onset of puberty is correlated with the height and weight of the individual, and the average height of boys has increased over the last century, and the change in the pitch of the voice of boys has advanced by one to two years.

The duration of the sexual cycles in mammals is related to that of the life span, to the size of the body and to the gestation period of the species; i.e. it is short in short-lived small animals with short periods of pregnancy. The latter correlation is modified in some species, such as the guinea-pigs, which produce more mature young after a prolonged period of gestation. Small animals with a short life span need a rapid rate of reproduction, hence they have short cycles and short gestation periods and periods of lactation. Even in them the actual period for conception is restricted to a short time of the cycle. This may not be different from the comparable period in larger animals, i.e. hours or at most days, but in the latter this period occurs less often because of the duration of the oestrous or menstrual cycles or the limited time of the breeding season, the length of the gestation period and of lactation.

Most of the facets of the sexual dimorphism of mammals become manifest at puberty, but may be due to modifications of the central mechanisms causing them. This is evident in the dual control of gonadotropins in females and the single one in males. Similarly the onset and duration of the growth spurt in males is due largely to the effect of androgens on the tissue cells, while in females it is caused mainly by the production of the pituitary growth hormone which is increased by oestrogens in parallel with the size of the hypophysis. Other hormones such as thyroxine and insulin are also involved in the promotion of growth. The elongation of the skeleton is limited by the ossification of the epiphyseal cartilages, which is speeded up more by oestrogens than by androgens. Hence the growth period in girls does not last as long as in boys. Defects in the production of oestrogens,

example may illustrate this point: the formation of red blood corpuscles in the bone marrow is influenced by the oxygen tension of the blood, the intake of iron, vitamin B12, folic acid, the intrinsic factor of the stomach and the erythropoietin formed in the kidney. The last compound responds to androgenic stimulation, and is more abundant in men and many male mammals, but the number of red blood corpuscles per millilitre of blood, though greater in men and many males, is equal in male and female rats, rabbits, goats and sheep. Men, bulls, stallions, male dogs, cats and hamsters have proportionately more red blood corpuscles than females, and castration reduces them, producing a mild anaemia which responds to androgenic medication; testosterone induces a polycythaemia (overproduction of red blood corpuscles) in spayed females. Thus in some species the production of erythrocytes depends more on the androgen-sensitive production of erythropoietin than in others.

Male mammals are more efficient than females in converting food into muscle, while females produce more fat. This is true even for rats, which utilise carbohydrates for their musculature and activity, and sheep which run on lipids. Detailed metabolic studies have been reported only for men, rats and mice, while for other species information is available for one sex only, either males or females. It is thus not possible to formulate the features common to all species in the sex differences of metabolic functions, though they exist in varying form. The reaction to drugs depends on enzyme activities, is influenced like the temperature regulation and basal metabolic rate by circadian and sexual cycles, and varies with species and sex.

The sexual dimorphism of mammals extends beyond the structure and function of the reproductive organs to the central regulations of all functions. The mosaicism of female somatic cells, and the uniformity as regards the X-chromosome of the male, influence the manifestation of inherited sex-linked disorders. The prevalence in one or the other sex of congenital abnormalities, sex-limited disorders, and of some infectious and of neoplastic diseases points to a basis in the normal physiological and structural organization of male and female mammals. The somewhat neglected study of this sexual dimorphism is of theoretical as well as of practical interest, not least in understanding the pathogenesis of disorders in men and animals and in finding means to prevent and cure them.

the complex somatic alterations take a long time to develop, and in boys as in girls of the same age appear first in the taller and heavier individuals. In animals environmental factors of nutrition, optic and olfactory stimulation, social conditions and presence of the other sex advance or delay the pubertal changes. The first oestrus and the first menses are easier to diagnose than the stage of spermiogenesis. While the changes in the reproductive organs at this period are of about equal duration in boys and girls, the activation of the more complex central regulatory activities in girls may require more time. During the last century the menarche has occurred at a decreasing age of girls. Whether the advance of pubertal changes in boys is similar is not well documented. It is likely, since the onset of puberty is correlated with the height and weight of the individual, and the average height of boys has increased over the last century, and the change in the pitch of the voice of boys has advanced by one to two years.

The duration of the sexual cycles in mammals is related to that of the life span, to the size of the body and to the gestation period of the species; i.e. it is short in short-lived small animals with short periods of pregnancy. The latter correlation is modified in some species, such as the guinea-pigs, which produce more mature young after a prolonged period of gestation. Small animals with a short life span need a rapid rate of reproduction, hence they have short cycles and short gestation periods and periods of lactation. Even in them the actual period for conception is restricted to a short time of the cycle. This may not be different from the comparable period in larger animals, i.e. hours or at most days, but in the latter this period occurs less often because of the duration of the oestrous or menstrual cycles or the limited time of the breeding season, the length of the gestation period and of lactation.

Most of the facets of the sexual dimorphism of mammals become manifest at puberty, but may be due to modifications of the central mechanisms causing them. This is evident in the dual control of gonadotropins in females and the single one in males. Similarly the onset and duration of the growth spurt in males is due largely to the effect of androgens on the tissue cells, while in females it is caused mainly by the production of the pituitary growth hormone which is increased by oestrogens in parallel with the size of the hypophysis. Other hormones such as thyroxine and insulin are also involved in the promotion of growth. The elongation of the skeleton is limited by the ossification of the epiphyseal cartilages, which is speeded up more by oestrogens than by androgens. Hence the growth period in girls does not last as long as in boys. Defects in the production of oestrogens, and with them failure of the growth of the hypophysis and secretion of growth hormone, inhibit the increase in height of such individuals compared with normal women, while the latter tend to be shorter than males on average, because the epiphyseal plates ossify earlier, and men are shorter than eunuchs, where the reduction in testosterone levels delays the ossification. Lack of androgens restricts the growth of castrate rats and bulls, but in both these species males are larger than females because of the stimulation of growth by androgens and the delay in ossification of the epiphyseal cartilage.

Though starting later, the growth spurt at puberty of boys lasts longer than in girls and contributes to the difference in height. Male rats continue to enlarge though at a decreasing rate throughout life, while females stop growing. The sex differences in size are not always in favour of males, since in some species the females are larger (e.g. rabbits, bats, golden hamsters). The reasons for the variation in different species of the relative size of males and females are obscure. A larger and more muscular male has an obvious advantage over his smaller rivals in the competition for a mate or a harem, and thus ensures his dominance in his society. Dominance does not necessarily go with size and is not important in species where the individuals live a solitary life except for the sporadic meeting, mating, and nurturing of the young. Size differences with sex occur in solitary as well as in gregarious species; the female golden hamster is larger than the male, but solitary, while the female spotted hyena is gregarious and larger than the male. How this greater growth of females is accomplished in these species is not clear. Dominant females tend to have higher levels of progesterone with its androgenic properties than have their female relatives, but this does not explain their advantage over the males.

The common features of sexual development of mammals are due to chromosomal, gonadal and neuroendocrine factors and are used in identical or at least similar manner by most species, until the adult body is formed subject to the genetic information specific for a species which is carried largely in the autosomes. Thus the sexual dimorphism of adult mammals is likely to vary more with the species and its adaptation to its environment. This leads to a different management of the available structures and functions; as outlined in previous chapters. The common features are thus reduced in number and obscured by the specific adaptations.

The wide variations in size, life span, nutritional requirements, choice of food, feeding and digestive periods, are reflected in the proportions of organs and their functions, as for instance the length and differentiation of the digestive tract in herbivores compared with that of carnivores, the specialization of the enzyme system required for the metabolism of the food and the excretion of the waste products. Sex differences are present in most organs of all mammals, but do not vary necessarily in the same direction, nor do the effects of gonadectomy or treatment with male or female sex hormones. The proportion to body weight of many internal organs from the heart to the kidney, the liver, the lung and the spleen, may be equal or greater in the females of some species and in the males of others. This applies even to the adrenal cortex, which is considerably larger in females of many species, but of the same relative size in hamsters of either sex. Only the larger female pituitary, the male gonad and possibly the female lympho-myeloid complex are consistent features in the sexual dimorphism. Male characteristics tend to be concentrated in the head and shoulder region, which mature earlier than those of the trunk and pelvic region. Antlers, horns, larger skulls and masticatory muscles, larger teeth and tusks, manes and beards, are distinguishing male features, while those of females are localized to the pelvis and ventral regions in the development of the mammary glands. The hair coat of males tends to be rougher than that of females and the sebaceous glands tend to be larger. Males tend to have more erythrocytes and females more white blood corpuscles per millilitre of blood in many, but not all, species. The amount of immunoglobulin IgM appears to be correlated with the number of X-chromosomes of the individual, and the successful transplantation of tissue is conditioned by histocompatibility factors on the Y-chromosome and of others on the maternal or paternal X-chromosome and their random inactivation.

The metabolism in all mammals shows circadian variations which are controlled by the brain through the pituitary and endocrine apparatus. The gonads feed their information into this system and in the case of females the sexual cycles are superimposed on the general biorhythms. Thus evidence for variations with the sexual cycle are found in the thyroid, adrenal, thymus, bronchus, the olfactory region, and possibly other organs, apart from the genital tract, the mammary and apocrine glands. These cycles are due in part directly to gonadal hormones, and partly indirectly to the generalized cyclical release of the hormones from the pituitary in response to the agents in the hypothalamus. All these hormones produce temporal modifications in the level of enzyme activities and metabolic functions, in addition to the circadian rhythms also present in the male.

The sensitivity of the enzyme systems to hormones varies with the species, and since many of them are employed to achieve a specific synthesis, they tend to vary with sex in comparable processes. An example may illustrate this point: the formation of red blood corpuscles in the bone marrow is influenced by the oxygen tension of the blood, the intake of iron, vitamin B12, folic acid, the intrinsic factor of the stomach and the erythropoietin formed in the kidney. The last compound responds to androgenic stimulation, and is more abundant in men and many male mammals, but the number of red blood corpuscles per millilitre of blood, though greater in men and many males, is equal in male and female rats, rabbits, goats and sheep. Men, bulls, stallions, male dogs, cats and hamsters have proportionately more red blood corpuscles than females, and castration reduces them, producing a mild anaemia which responds to androgenic medication; testosterone induces a polycythaemia (overproduction of red blood corpuscles) in spayed females. Thus in some species the production of erythrocytes depends more on the androgen-sensitive production of erythropoietin than in others.

Male mammals are more efficient than females in converting food into muscle, while females produce more fat. This is true even for rats, which utilise carbohydrates for their musculature and activity, and sheep which run on lipids. Detailed metabolic studies have been reported only for men, rats and mice, while for other species information is available for one sex only, either males or females. It is thus not possible to formulate the features common to all species in the sex differences of metabolic functions, though they exist in varying form. The reaction to drugs depends on enzyme activities, is influenced like the temperature regulation and basal metabolic rate by circadian and sexual cycles, and varies with species and sex.

The sexual dimorphism of mammals extends beyond the structure and function of the reproductive organs to the central regulations of all functions. The mosaicism of female somatic cells, and the uniformity as regards the X-chromosome of the male, influence the manifestation of inherited sex-linked disorders. The prevalence in one or the other sex of congenital abnormalities, sex-limited disorders, and of some infectious and of neoplastic diseases points to a basis in the normal physiological and structural organization of male and female mammals. The somewhat neglected study of this sexual dimorphism is of theoretical as well as of practical interest, not least in understanding the pathogenesis of disorders in men and animals and in finding means to prevent and cure them.

Abbreviations and glossary

A mice: an inbred strain of mice.
A-cells: cells of islets of Langerhans producing glucagon.
Acromegaly: a disorder with enlargement of the face, hands and feet.
Acrosome: a cap on the nuclear part of the head of the spermatozoon.
ACTH: adrenocorticotropin.
Adenoma: a benign tumour of glandular tissue.
ADH: antidiuretic hormone.
Adrenalin (epinephrine): formed in the medulla of the adrenal.
Adrenergic nerves: release adrenalin or nor-adrenalin at their ends.
Adrenocorticotropin (ACTH): a hormone secreted by the anterior pituitary which regulates the production and release of steroids in the adrenal cortex.
Agammaglobulinaemia: a deficiency of gamma globulins which lowers the resistance to infections.
Aldosterone: a mineralcorticoid produced in the outer zones of the adrenal cortex; involved in the regulation of the mineral metabolism.
Alopecia: baldness.
Anagen: the growth phase of the hair cycle.
Anastomosis: communication between blood vessels.
Angiotensin: a polypeptide regulating blood pressure (cf. renin).
Ankylosing spondylitis: stiffening of the spinal column by bone formation between the vertebrae.
Ankylosis: immobilization of joints.
Antiandrogens: substances which counteract the effects of androgens.
Antidiuretic hormone (ADH, vasopressin): produced in the posterior pituitary; it acts on the kidney to regulate the water and salt balance.
Apocrine glands: these shed part of their cytoplasm in their secreta.
APUD system: amine precursor uptake and decarboxylase system—peptides with endocrine or paracrine functions, produced in the nervous system and gastrointestinal tract.
Atherosclerosis: deposition of lipids in and hardening of the arterial walls.

Autosomes: chromosomes other than the sex chromosomes.

BALB mice: an inbred strain of mice.
Barr body: a small body of heterochromatin in the interphase nucleus of female cells representing the heteropycnotic X-chromosome. (cf. drum-stick).
B-lymphocytes: a group of lymphocytes responsible for humoral immune reactions and the production of immunoglobulins. (cf. T-lymphocytes)

Cachexy (adjective cachectic): wasting in patients caused by their disease.
Calcitonin: a hormone of the thyroid gland; it helps to regulate the mineral metabolism.
cAMP: cyclic adenosine monophosphate.
Carcinogen: a cancer inducing agent.
Catecholamines: amines including epinephrine, nor-epinephrine, dopamine; they are synthesized by cells derived from the neural crest.
CBA mice: an inbred strain of mice.
C57Bl, C57Bl/6: inbred strains of mice.
Chiasma: the crossing-over of segments between homologous chromosomes in meiosis.
Cholecystokinin (pancreozymin): a duodenal hormone which stimulates the secretion of enzymes in the pancreas and the contraction of the gall-bladder.
Cholinergic nerves: these release acetylcholine at their ends.
Choriocarcinoma: a carcinoma derived from chorionic epithelium.
Chromaffine tissue: stains with chromium salts; it is derived from the neural crest.
Colchicine: the alkaloid of *Colchicum autumnale*.
Corticosteroids: the mineralcorticoid aldosterone and the glucocorticoids cortisone, hydrocortisone and their precursors and metabolites produced in the inner zones of the adrenal cortex.
Corticotropes: cells in the anterior pituitary, secreting ACTH.
Cryptorchism: failure of the testis to descend into the scrotum.
Cytokinesis: division of the cytoplasm in mitosis.

DA: dopamine.
Desoxyribonucleic acid (DNA): D in haploid amount, 2D in diploid and 4D in tetraploid amount.
Df: differentiating, postmitotic cell.

Dg: degenerating cell.
Diakinesis: the final stage of the prophase in the first meiotic division.
Diploid cells: cells containing pairs (2n) of homologous chromosomes.
DNA: desoxyribonucleic acid.
Dopamine (DA): a catecholamine and precursor of nor-epinephrine.
Drum-stick: a heterochromatic body of the nucleus of female leukocytes, representing the heteropycnotic X-chromosome. (cf. Barr body).
Dyslexia: reading disability.

Ectopic: out of place.
Enteroglucagon: a glucagon produced in the intestine.
Epinephrine: adrenalin.
Erythropoietin: produced in the juxtaglomerular apparatus of the kidney and involved in the formation of red blood corpuscles.
Eukaryocytes: cells in which the genome is confined to the nucleus and separated from the cytoplasm. (cf. prokaryocytes).
Eutherian: mammalian species which form a placenta in pregnancy.

Foetoplacental unit: it is formed by the foetal adrenal and the placenta.
5-HT: 5-hydroxytryptamine.
5-hydroxytryptamine (5-HT, serotonin): a neurotransmitter.
5-hydroxytryptophane: a precursor of 5-HT.
Freemartin: the sterile female of a bisexual pair of twins in ruminants.
FSH: follicle stimulating hormone produced in the anterior pituitary.

Gametes: germ cells of the series spermatogonia to spermatozoa and oogonia to ova.
Gastrin: stimulates the gastric secretion, produced in the pyloric region.
Genome: the set of chromosomes.
Glucagon: produced in the islets of Langerhans; it regulates the level of blood sugar, and inhibits the secretion of acid in the stomach.
Glucocorticoids: cf. corticosteroids.
Goitrogens: substances which interfere with the normal functions of the thyroid and lead to the formation of goitres.
Gonad: a gland which produces germ cells and hormones such as androgens, oestrogens, progesterones.
Gonadotropes: cells in the anterior pituitary which secrete FSH and LH or ICSH.
G1, G2, Go: phases in the cell cycle: G1 precedes the synthesis of

DNA (S), G2 follows it and precedes mitosis (M), Go is the term for a cell of the reproductive compartment which is not cycling, but capable of doing so and may function as an intermitotic differentiating cell.

Growth hormone (somatotropin): produced in the anterior pituitary.

Gynaecomasty: an enlargement of the male mammary glands.

Haemopoietic tissues: tissues involved in the formation of blood.

Haploid cells: cells having a single set (n) of unpaired chromosomes. (cf. diploid).

HCG: human chorionic gonadotropin.

Hepatoblastoma: a tumour derived from liver cells.

Hepatotoxic substances: substances injurious to the liver.

Heterogametic: the sex with two dissimilar sex chromosomes: XY in male mammals and WZ in female birds.

Heterokaryons: nuclei formed by the fusion of cells originating in different tissues or species.

Heteromorphic: differing in appearance.

Heteropycnosis: the differential condensation of parts of or of whole chromosomes.

Heterozygotes: have differing alleles on corresponding loci of a homologous pair of chromosomes.

HGPRT: hypoxanthine-guanine phosphoribosyl transferase.

Hiatus hernia: protrusion of parts of the stomach through the diaphragm.

Histocompatibility factors: genes determining the compatibility with one another of tissues and cells of two individuals.

HIOMT: hydroxy-indol-*O*-methyltransferase; an enzyme involved in the formation of melatonin from serotonin.

Holocrine glands: these shed whole cells as secreta (cf. apocrine).

Homogametic: the sex with a pair of similar sex chromosomes: XX in female mammals, WW in male birds.

Homozygotes: these have the same alleles on corresponding loci of a homologous pair of chromosomes.

HY-antigens: histocompatibility factors carried on the Y-chromosome.

Hydrocephalus: a skull enlarged by increased amounts of cerebrospinal fluid.

Hypohidrotic: producing reduced quantities of sweat.

Hypospadias: failure of the urethra, clitoris or penis to close.

Hypothalamic releasing (RF) and inhibiting factors (IF): factors acting on the pituitary (e.g. LRF: releasing factor for gonadotropins, PIF: prolactin inhibiting factor).

ICSH: interstitial-cell-stimulating hormone.
Ig: immunoglobulins appearing as IgA, IgD, IgE, IgG or IgM.
Inhibin: an agent formed in the seminiferous tubules which regulates the production of spermatozoa.
Insulin: produced by the B-cells of the islets of Langerhans; one of its main functions is the control of the metabolism of sugar.
Interstitial-cell-stimulating hormone (ICSH): the hormone produced by the anterior pituitary which regulates the activity of the interstitial cells of the testis (identical with LH).

Kallikrein: an enzyme, which, like amylase, is secreted by the acinar cells of the submaxillary gland of mice.
Karyokinesis: division of the nucleus in mitosis. (cf. cytokinesis).
Kleinfelter's syndrome: agonadal dysgenesis due to a XXY constitution.

LH: luteinizing hormone.
Lipolysis: mobilization of fat.
Lordosis: forward curvature of the spinal column.
LRF: a releasing factor produced in the hypothalamus, which acts on the gonadotropes.
Luteinizing hormone (LH): produced in the anterior pituitary and causing the formation of corpora lutea in the ovary (identical with ICSH).

Medullarin (factor X): produced by the foetal testis; acts on the differentiation of the genital tract.
Meiosis: reduction divisions leading to the haploidy of germ cells.
Melanocyte stimulating hormone (MSH): hormone of the anterior pituitary.
Melatonin: a hormone produced by the pineal body from serotonin. (cf. HIOMT).
Menarche: the onset of menstrual periods in the puberty of females.
Mesonephros (Wolffian body): the second stage in the development of the kidney in vertebrates.
Mice: strains A, BALB, C57Bl, CBA, Tfm and W/W are obtained and continued by inbreeding through brother-sister mating or through back-crossing with one of the parents.
Monoestrous: having a single oestrous period during the breeding season.
Monosomy: the chromosome complement has only one of a homologous pair; e.g. XO instead of XX.

MSH: melanocyte stimulating hormone.

Nerve growth factor: an agent produced in the secretory tubules of the submaxillary gland of male mice.
Neuroendocrine cells: ganglion cells which discharge their specific secretions into the blood, and act on target cells in the anterior and posterior pituitary.
Neuropil: the mass of interwoven dendrites and axons within the body of ganglion cells.
Nude mice: a strain without a thymus.

Osteoarthrosis: a disease leading to immobilization of joints.
Osteomalacia: softening and deformation of bones due to a depletion of the mineral content.
Osteoporosis: rarefaction of the bones.
Otosclerosis: ankylosis of the ossicles in the middle ear.
Oxytocin: a hormone produced in the posterior pituitary which stimulates the myoepithelial cells of the mammary glands and the smooth muscles of the uterus.

Pachytene: a stage in the prophase of the first meiotic division.
Pancreozymin: cholecystokinin.
Parabiosis: the joining of two individuals to establish a common blood circulation.
Paracrine agents: these act at short range on neighbouring cells and tissues.
Parathormone: a hormone produced by the parathyroids, involved in the regulation of the mineral metabolism.
Parenchyma: the specific cells of an organ.
Perinatal: the period around birth.
Phenotype: the characteristics manifested by an individual.
Pheromones: these serve as olfactory signals for other members of the same species.
Polygenic: an effect produced by the combined action of numerous genes.
PRL: prolactin.
Prokaryocytes: organisms without nuclei and thus without separation of the genetic and cytoplasmic material; they are haploid and propagate asexually.
Prolactin (PRL): a hormone produced by the anterior pituitary which acts on the mammary gland.
Prooestrus: the stage in the cycle preceding oestrus.

Abbreviations and glossary

Prostaglandins: these are produced by many tissues from fatty acids and serve diverse functions.

Recessive genes: these are manifest in a homozygous, but not in a heterozygous, individual.
Regaud's residual bodies: cytoplasmic remains discarded during spermatogenesis.
Renin: a proteolytic enzyme secreted by the kidney and involved in the formation of angiotensin.

Scoliosis: a torsion of the spinal column.
Secretin: a substance produced by the intestinal mucosa, the first hormone discovered; it stimulates enzyme secretion of the pancreas.
Serotonin: 5-HT.
Somatotropin: growth hormone.
S phase: the stage of DNA synthesis in the reproductive cell cycle.

Teleangiectasy: dilatation of capillaries and small arteries.
Telogen: the resting phase at the end of the hair cycle.
Teratogens: substances causing developmental abnormalities in embryos.
Teratoma: a tumour composed of foetal tissues.
Tetraploid cells: cells with two sets (4n) of paired chromosomes.
Thymosin: a thymus extract which increases the number of lymphocytes and the immune reactions to skin grafts in newborn mice deprived of a thymus.
Thyrotropes: cells of the anterior pituitary which produce TSH.
T-lymphocytes: a group of lymphocytes responsible for the cellular immune reactions and thymus-dependent.
Trisomy: the presence in triplicate of one chromosome e.g. trisomy 21, Down's disease. (cf. monosomy).
TSH: thyroid stimulating hormone.
Turner's syndrome: gonadal dysgenesis due to an XO constitution.

Vasopressin: ADH.

Wolffian body: mesonephros.

Prostaglandins: these are produced by many tissues from fatty acids and serve diverse functions.

Recessive genes: these are manifest in a homozygous, but not in a heterozygous, individual.
Regaud's residual bodies: cytoplasmic remains discarded during spermatogenesis.
Renin: a proteolytic enzyme secreted by the kidney and involved in the formation of angiotensin.

Scoliosis: a torsion of the spinal column.
Secretin: a substance produced by the intestinal mucosa, the first hormone discovered; it stimulates enzyme secretion of the pancreas.
Serotonin: 5-HT.
Somatotropin: growth hormone.
S phase: the stage of DNA synthesis in the reproductive cell cycle.

Teleangiectasy: dilatation of capillaries and small arteries.
Telogen: the resting phase at the end of the hair cycle.
Teratogens: substances causing developmental abnormalities in embryos.
Teratoma: a tumour composed of foetal tissues.
Tetraploid cells: cells with two sets (4n) of paired chromosomes.
Thymosin: a thymus extract which increases the number of lymphocytes and the immune reactions to skin grafts in newborn mice deprived of a thymus.
Thyrotropes: cells of the anterior pituitary which produce TSH.
T-lymphocytes: a group of lymphocytes responsible for the cellular immune reactions and thymus-dependent.
Trisomy: the presence in triplicate of one chromosome e.g. trisomy 21, Down's disease. (cf. monosomy).
TSH: thyroid stimulating hormone.
Turner's syndrome: gonadal dysgenesis due to an XO constitution.

Vasopressin: ADH.

Wolffian body: mesonephros.

Reading list and reference books

Altman, P. L. and Dittmer, D. S. (1972). *Biology Data Book*, second edition.
  (1962). *Growth*
  (1968). *Metabolism*
  Washington D.C.: Federation of American Societies of Experimental Biology.
Bach, F. H. and Good, R. A. (1972). *Clinical Immunobiology*. New York and London: Academic Press.
Berg, R. T. and Butterfield, R. M. (1976). *New Concepts of Cattle Growth*. Sydney University Press.
British Medical Bulletin (1969). New aspects of human genetics, Vol. 25, part 1.
  (1970). Control of human fertility, Vol. 26, part 1.
  (1971). Epidemiology of non-communicable diseases, Vol. 27, part 1.
  (1974). Development and regeneration in the nervous system. Vol. 30, part 2.
  (1976). Human malformations, Vol. 32, part 1.
  (1976). Immunological tolerance, Vol. 32, part 2.
Cole, H. A. and Cupps, P. T. (1969). *Reproduction in Domestic Animals*. New York and London: Academic Press.
Cove, D. J. (1971). *Genetics*. Cambridge University Press.
Dillon, R. S. (1973). *Handbook of Endocrinology*. Philadelphia: Lea & Febiger.
Duncan, R. and Weston-Smith, M. (1977). *The Encyclopaedia of Ignorance*, Vol. 2 (articles by R. J. Britten, F. H. C. Crick, A. E. Mourant, and J. Maynard Smith). Oxford: Pergamon Press.
Gasser, D. L. and Silvers, W. K. (1972). Genetics and immunology of sex-linked antigens. *Advances in Immunology*, 15, 215-247
  (1974). Genetic determinants of immunological responsiveness. *Advances in Immunology*, 18, 1-66.
Glucksmann, A. (1974). Sexual dimorphism in mammals. *Biological Reviews*, 49, 423-475.
Good, R. A. and Fisher, D. W. (1974). *Immunobiology*. Sunderland, Mass.: Sinauer Associates.
ter Haar, M. B. (1976). Circadian rhythms of protein metabolism in the rat brain. *Progress in Animal Biometeorology,* Vol. 1, part 2. Lisse, The Netherlands: Swets & Zeitlinger.
Hafez, E. S. E. (1968). *Adaptation of Domestic Animals*. Philadelphia: Lea & Febiger.
Lewis, K. R. and John, B. (1968). The chromosomal basis of sex determination. *International Review of Cytology*, 23, 277-379.

Litwack, G. (1970). *Biochemical Actions of Hormones*, Vol. 1 (see article by B. Black and J. Axelrod) New York and London: Academic Press.

Lodge, G. A. and Lamming, G. E. (1968). *Growth and Development of Mammals*. London: Butterworth.

Lyon, M. F. (1972). X-chromosome inactivation and developmental patterns in mammals. *Biological Reviews*, 47, 1-35.

McKusick, V. A. (1962). On the X-chromosome in man. *Quarterly Review of Biology*, 37, 69-175.

(1975). *Mendelian Inheritance in Man*. Johns Hopkins University Press, Baltimore.

Martini, L. and Ganong, W. F. (1966). *Neuroendocrinology*, Vols. 1 and 2. New York and London: Academic Press.

Melmon, K. L. and Morelli, H. F. (1972). *Clinical Pharmacology*. New York: The Macmillan Co.

Mittwoch, U. (1973). *Genetics of Sex Differentiation*. New York and London: Academic Press.

Money, J. and Ehrhardt, A. E. (1972). *Man and Woman, Boy and Girl*. Baltimore: Johns Hopkins University Press.

Monteith, J. L. and Mount, L. E. (1974). *Heat Loss from Animals and Man*. London: Butterworth.

Montgomery, D. A. D., and Welbourn, R. B. (1975). *Medical and Surgical Endocrinology*. London: Edward Arnold.

Ohno, S. (1967). *Sex Chromosomes and Sex Linked Genes*. Berlin, Heidelberg and New York: Springer Verlag.

(1970). *Evolution by Gene Duplication*. Berlin, Heidelberg and New York: Springer Verlag.

Ounsted, C. and Taylor, D. C. (1972). *Gender Differences: their Ontogeny and Significance*. Edinburgh: Churchill-Livingstone.

Ringertz, N. R. and Savage, R. E. (1976). *Cell Hybrids*. London and New York: Academic Press.

Ruch, T. C. and Patton, H. D. (1974). *Physiology and Biophysics*. Philadelphia, London, Toronto: W. B. Saunders.

Solari, A. J. (1974). The behaviour of the XY pair in mammals. *International Review of Cytology*, 38, 273-318.

Svenson, M. J. (1970). *Dukes' Physiology of Domestic Animals*, 8th edition. Ithaca and London: Cornell University Press.

Index

Abnormality congenital 9, 65, 67, 119, 158, 162
    sex-linked cf X-chromosome 10, 68, 154
Abomasum 103
Abortions 60, 158
Acetaldehyde 142
Acetylcholine 111
Acid production (stomach) 145
Acromegaly 81, 126
Acrosome 29
ACTH cf pituitary 40, 108, 109, 115, 121, 123, 127, 128, 133, 140
Adenohypophysis cf pituitary 122
ADH 40, 104, 124, 129
Adipose tissue cf fat 42, 75, 80, 92-94, 97, 98-100, 136, 144
Adrenal 35, 38, 46-48, 98-100, 108, 126-128, 138, 139, 161
    in foeto-placental unit 35, 37-39, 46, 51, 52, 54-56
Adrenalin cf epinephrine 111, 121, 127
Adrenergic nerves 111
Afterbirth 48
Agammaglobulinaemia 15
Alcohol 111
Aldosterone 119, 147
Alkaline pH 62
Alopecia (baldness) 95, 137, 138
Anaemia 162
Anastomoses 33
Androgens, production 31, 36, 45, 55, 109, 122, 127, 156, 159
    in adrenals 127, 128
    in ovary 128
    effects 42, 46, 68, 80-82, 108, 128, 130-134, 136, 138, 146, 155, 158, 162
Androstenedione 127
Anencephaly 66, 67
Angiotensin 119, 127, 147
Ankylosis 91
Anoestrus 44
Anovulatory cycle 49
Antiandrogens 34, 42, 45, 46, 108, 139, 140, 147

Antibodies 70, 125
Antigens (HY) 60
Antisera 61
Antlers 54, 89, 94, 140
Aorta 67
Apocrine glands 96, 139, 161
APUD system 47, 111, 113, 121
Arctic mammals 139
Arcuate nucleus 132, 133
Aspirin 112
Atherosclerosis 119
Atrophy of genital organs 44
Audiogenic seizures 135
Auditory stimuli 134
Autoimmune disorders 70, 118, 119
Autosomes cf chromosomes 5, 12-18, 23, 153, 157, 160
Axillary glands and hairs 137-139
Azodye 144

Bacteria 3, 71
Barbiturates 107, 112, 125, 144
Barr body cf X-chromosome 11, 59, 153
Biological clock 44, 53, 109, 110
    synchronizers 110
Biorhythms, circadian and sexual 102-113, 131-134
    drug sensitivity 111-113
Birds 5, 6
Bladder 147
Blocking agents 131
Blood 114, 115
    anaemia 162
    blast cells 116
    erythrocytes 97, 114, 147, 161, 162
    granulocytes 115, 161
    haemoglobin 114
    lymphocytes (B and T) 115-117
    menstrual bleeding 115
    plasma 114,
    platelets 115
    polycythaemia 115, 162
    pressure 119
    sugar 81, 109, 122, 126
Bone 90, 91

Bone in clitoris 50, 91
  marrow 114, 115, 119, 162
Bowel transit times 103
Bowman's capsule 97, 146
Brachial glands 140
Brain cf hypothalamus, pineal body and pituitary 97-101, 110, 130-135
Breast (mammary glands) 33, 45, 96, 133, 137, 140, 161
  cancer 70, 71
Breeding seasons 53, 54, 96, 105, 124, 127, 131, 137, 139, 141, 156

Cachexia 92
Caecum 103
Calcitonin 121, 122, 125, 147
Calcium 125, 147
Cancer 64, 70, 111
Canines 143
Capacitation 55, 157
Carnivores 103, 160
Castration 35, 42, 45, 68, 80, 91-93
Castration cells 123
Catecholamines 94, 127
Cell cycle 20-22, 72-74, 83, 84
  biorhythms 22, 132, 153
  cytokinesis 24
  duration 21
  generations 82-84
Cell loss 22, 62, 74
Cell proliferation 22, 62, 75
Chiasma cf meiosis 94
Childbirth 64
Chimaera 34
Chloroform 112
Cholecystokinin 121
Cholesterol 34, 42, 119
Cholinergic nerves 111
Chorionic gonadotropin 54
Chromaffin tissue 127
Chromosomes 3-6, 8-18, 68, 153, 158
    abnormalities 68, 89, 128, 158
    autosomes 5, 12-18, 23, 153, 157, 160
    condensed regions 13
    genetic analysis 9
    hetero-, homo-zygote 11
    labelling 13
    monosomy 12, 14, 16, 153
    morphological identification 8
    mutation 4, 158
    non-disjunction 28, 154
    number 4, 8, 12, 153
    number of loci 10, 12
    pairing at meiosis 24
    ploidy 4
    separation in meiosis 19
    translocations 28, 154, 158

  sex chromosomes 5, 8, 12-18, 23, 30, 41, 89, 153
    hetero-, homo-gametic 5, 6
    unusual combinations 16, 154
    X-chromosome 5, 6, 12-18, 153-155, 157
      abnormalities linked to it 10, 68, 154
        agammaglobulinaemia 15
        colour blindness 10, 68, 154
        haemophilia 10, 68, 154
        HGPRT 15, 18, 155
        hypohidrotic dysplasia 15, 139
        immunodeficiency 68
        mental deficiency 10, 68
      Barr body 11, 59, 153
      dosage compensation 14, 153
      drum-stick 11, 153
      duplication 16
      heteropycnosis 11, 14-18, 24, 138, 154
      histocompatibility and IgM 117, 118
      mosaicism in somatic cells 14, 15, 17, 138, 139, 154, 162
    Y-chromosome 5, 6, 8, 14, 16, 17, 131, 153-155
      genes 13, 14, 23
      histocompatibility factors 13, 23, 118, 119
      meiosis 23-28
      repeated DNA sequences 24
      Y-body, stains 61
Circadian rhythms 102, 103, 108-110
  endocrines 122, 161
  free-running periods 108
  metabolism 105-112
  photoperiod 102, 103, 108-110
  pineal gland 110, 124
  sexual cycles 103, 106, 107, 153
  synchronizers 110
  temperature 103
Cleft lip and palate 66, 67
Club foot 66, 67
Colchicine 8
Collagen 94, 97
Colour 94, 95
Concanavalin 116
Conception 159
Conformation in cattle 92
Conjugation 3
Copulation 55, 157
Corpus luteum 43, 44, 54
  striatum 131
Corticosteroids 42, 121, 122, 127, 140
Corticotropes 40
Cortisone 66, 109

Index

Creatine 81
Cretinism 125
Cryptorchism 38
Cyclic AMP 94

Deformations 66, 158
Degeneration of cells 62, 74
Dehydrotestosterone 35
Dentition 143
Dermis 94, 136
Diabetes mellitus 112, 126
Diaphragm 142
Differentiation 21, 73, 74, 158
Digestion 103
Digestive tract 142-145, 160
Dislocation of hip 66, 67
DNA 3
  hybridization 9
  labelling 13
  multiple copies 12, 24
  synthesis 8, 13, 19, 20, 21, 24-27, 83, 109
Dominant sex 131, 139
Dopamine cf neurotransmitters 110, 111, 130, 131
Dormancy of fertilized ova 152
Down's syndrome 16, 28, 56, 66, 67, 157
Drum-stick 11, 153
Ductus arteriosus 67
  deferens 31
Dwarfism 82
Dyslexia 135

Elastic fibres 94
Emphysema 142
Endocrine system 121-129
Endoderm 30
Endometrium 43,
Endomorphine 131
Enkephalin 131
Enteroglucagon 109, 121
Enzymes 3, 28, 102, 116, 145, 161, 162
  α-reductase 35, 44
  amylase 143
  dehydrogenase 143
  5 β-enzymes 143
  glucuronidase 147
  HGPRT 15, 18
  HIOMT 110
  hydrosteroid dehydrogenase 143
  hydroxylase 112
  kallikrein 143
  oxidative enzymes 132
  phosphatase 144, 147
  proteolytic enzymes 143
  20-oxoreductase 143
  tyrosine transaminase 112

Epididymis 30, 32, 33, 95, 136, 157
Epinephrine-adrenalin 111, 121, 127
Epiphysis 75, 84, 159
Ergot 112
Erythrocytes cf blood 97, 114, 147, 161, 162
Erythropoietin 97, 114, 122, 147, 162
Escherichia coli 12, 13
  endotoxin 105, 111
Ethanol 142, 144
Ethionine 144
Ethylene glycol 112
Eukaryocytes 4, 13
Eunuchs 69, 97
Eutherian mammals 6, 154
Eyes 66

Factor X = medullarin 34, 36, 47, 155
Fat = adipose tissue 42, 75, 80, 92-94, 97, 98-100, 136, 144
  brown 94, 104
  colour and composition 93, 94
Feeding times 103
Feminization 128
Fertilization 25, 51, 59, 151, 152, 157
Fibroblasts 83, 155
5HT cf neurotransmitters 104, 108, 110, 111, 131, 132
Fingerprints 136
Flank glands and organs 96, 140
Fleece 94
Flora and fauna 103
Foeto-placental unit 35, 37-39, 46, 51, 52, 54-56
Follicles of ovary 49, 51, 56, 128
Follicular phase 43, 51
Freckles 95, 137
Freemartins 33, 34, 46, 89, 155
Free-running period 108, 109
FSH 40, 43, 51, 109, 122, 123, 133, 157

Gall Bladder 121, 144, 145
Gastric ulcer 145
Gastrin 109, 121, 126
Genes
  alleles 10
  colour 95
  dominant 13
  histocompatibility 9, 13, 23, 117, 118, 119
  integrator 13
  linkage maps 10
  mutant: h 10
    T locus in mice 62
    Tfm in mice 36
    W/W in mice 31
  mutation 3, 10

number 12, 153
polygenic action 69
reactivation 13
recessive 10
regulatory 12, 13
sensor 13
structural 12, 153
German measles 65
Gestation-pregnancy 39, 52, 76, 82, 131, 133, 137, 156, 159
Glucagon 94, 122, 126
Glycogen 81, 144
Glycoproteins 37, 121
Goitre, goitrogens 125, 138
Gonadotropes 40, 122, 123
Gonadotropins 40, 42, 43, 46, 123, 124, 129, 132, 133
Gonads 96-98, 121, 124, 128, 139
   development 30-38
   genital ridge 30
   Mullerian ducts and derivatives, 30-33, 155
   Wolffian ducts and derivatives 30-33
   secretory activity 6, 7, 31, 33-38
   gonocytes, migration, multiplication 30, 31, 36, 37
   gametogenesis in ovary 6, 26-28, 48-50, 97, 98
   in testis 6, 27, 28, 48, 50-52, 54-56, 97, 98
   effects on differentiation of:
      androgens 33, 34, 42, 45, 46, 50
      antiandrogens 34, 42, 45, 46, 50
      gonadectomy 34, 41-45, 80, 91-93, 101, 117, 123, 124, 128, 134, 138, 139, 161
      HCG 37, 38
      hypothalamus 39-45
      oestrogens 42, 45, 46, 50
      pineal body 40-42, 49, 50
      pituitary 40-45, 133
      puberty 40, 42, 48-55, 159
      seasonal breeding activity 53, 54, 96, 105, 124, 127, 131, 137, 139, 141, 159
      sexual cycles 40-45, 50-53, 95, 97, 122, 131, 134, 140, 152, 153
Gout 69, 112
Grafts of gonads 41, 44
   hypothalamus 41, 45
   pituitary 41, 45, 132
Granulocytes 115
Granulosa cells 128
Growth 72-85, 158
   effect of endocrines 80-85
   puberty 78-82, 100, 158

   epidermal and nerve factor 97, 122, 143
Growth hormone 40, 81, 94, 109, 121, 123, 133, 159
Gynaecomasty 144

Haemoglobin 114
Haemophilia 10, 66, 68, 154
Haemopoietic tissue 114
Hair coat 94, 96, 136-139, 161
   colour 15, 17, 94, 95, 139
   cycle 74, 84, 97, 136-138
   types 50, 80, 82, 84, 89, 94
Hashimoto's thyroiditis 118
Hassal's corpuscles 116
HCG 37
Heart 98-100, 146
   defects 67, 119
   diseases 64, 69
   rate 104
Hepatotoxic agents 143-145
Herbivores 160
Hermaphrodite 4, 5
Heterogametic 5, 6, 154
Heterografts 116
Heterokaryons 9
Heteropycnosis 11, 14, 154
Heterozygous 11, 158
HGPRT 15, 18
Hiatus hernia 145
Hide 94
Hip joint 90
Hirsutism 127
Histocompatibility 9, 118, 119, 161
Holocrine glands 139
Homogametic 5, 6, 154
Homozygous 11, 158
Hormones 121-129
   protein binding 36, 122, 128
   receptors 36, 121, 122, 132
   secretion in tumours 128, 129
Hybridization of DNA 9
Hydrocephalus 67
HY — antigens and antisera 60, 61
Hypertension 119, 147
Hypohidrotic dysplasia 15, 139
Hypophysis cf pituitary
Hypospadias 45, 50
Hypothalamus 39, 41, 100, 101, 108, 111, 130, 132-135, 139
   controlling factors 40, 45, 80, 132-135, 156
   critical phase 42, 44-46, 132, 155

ICSH 52
Implantation of ova 60

Index

Immune system 115-119
  autoimmune diseases 118, 119, 125
  deficiencies 68, 70, 118
  hyperimmunization 119
  immunoglobulins 116, 117
    X-linked IgM 117, 161
  reactions: cell mediated 115
    humoral 115, 119, 142
Incisors 143
Inherited characters 10
  sex-linked and sex-limited 11
Inhibin 52
Insemination 61, 62
Insulin 94, 109, 122, 126, 129, 159
Intersex 34, 35
Interstitial cells of ovary 37
  testis 31, 34, 36, 52, 155
Intestinal cells 72, 84, 103, 109, 121
Invariance 3, 23
Iodine 85, 106, 107, 124, 125
Iron 162

Juxta-glomerular apparatus 97, 115, 122, 147
Juxta-medullary X-zone 127

Kangaroo 16, 49
Karyokinesis 19
Kidney 97-100, 146, 147

Lactation 45, 54, 78, 109, 131, 133, 140, 152
Larynx 141
Length 75
Leukaemia 67
Leydig cells 31, 34, 36, 52, 155
LH 43, 51, 52, 55, 56, 107, 109, 111, 122-124, 133, 157
Life span of mammals 58, 64-71, 82-85, 159
  effect of gonadectomy 68, 69
    oestrogens 68
    testosterone 68
    weight 68, 82
Life span of cells 74, 83, 97, 152
Linkage maps 10
Lipase 94
Lipids 162
Lipolysis 94
Lipoproteins 69, 119, 120
Liver 39, 44, 49, 98-100, 112, 121, 125, 127, 140, 142-144
Locomotory apparatus 97
Lordosis 43, 51
Lung 98-100, 133, 141, 142, 146
  cancer 70, 128, 129, 142

Luteal phase 43, 51, 56
Lymphocytes 114-119

Macaca speciosa 95, 138
Macrophage 116
Malaria 64
Malformations 66
Malnutrition 64
Mammary glands 33, 45, 96, 133, 137, 140, 161
Marsupials 152, 154
Maternal age 56, 157
Maternal factors 62, 76
Maturation, rate 72-85, 130
Medullarin 34, 36, 47, 155
Megakaryocytes 115
Meiosis 23-29, 156, 167
Melanocytes 136
Melatonin 4, 41, 108, 110, 124
Menarche 49
Meningitis 69
Menopause 48, 55, 69, 115
Menstrual cycle 22, 52, 53, 95, 97, 115, 122, 134, 139-141, 144, 152
Metabolism:
  biorhythms 102, 113
  digestive tract 103
  hormones 80-82
  rate 97, 102-113, 161
Mesonephros 30-33, 155
Methionine 105, 132
Methylthiouracil 138
Microtus oregoni 16
Minerals 147
Mitosis 19-22, 72-74, 83, 139
Mononucleate cells 21
Monoestrous 53, 54
Monophasic 103
Monosomy 12, 14, 16, 153
Monotremes 6
Morphine 112
Mouse strains:
  A 62, 119
  absence of eyes 66
  alkaline pH 62
  BALB 117, 118
  CBA 117
  C57B1 6
  nude 116, 126
  Tfm 36
  T-locus 62
  W/W 31
MSH 123, 136, 137
Mucus 141
Mullerian duct 30-33, 155
Multinucleate cells 21

Muscles 91, 92
Musk 134
Mutation 3, 158
Myopus schisticolor 16, 154

Nembutal 107
Nephrectomy 146
Neuroendocrines (hypothalamus, pineal body, posterior pituitary) 39-47, 130-135, 156
Neuropil 132
Neurotransmitter substances 40, 110-113, 133
   acetyl choline 111
   adrenalin 111, 121, 127
   dopamine 110, 111, 130, 131
   5 HT(5-hydroxytryptamine, serotonin) 104, 108, 110. 111, 130, 131, 132
   norepinephrine 105, 110, 111, 127

Obesity 93
Oesophagus 145
Oestrogen 7, 38, 42, 43, 45, 49, 69, 81, 122, 127, 128, 139, 144, 155, 159
Oestrous cycle 22, 41, 42, 51, 53, 105-108, 115, 117, 122, 133, 153, 159
Oestrus 44, 51, 152
Olfactory bulb 131, 134
Olfactory epithelium 134, 141
Oogenesis 27, 156
Organs, proportion to body 96-101
Osteoarthrosis 91
Osteomalacia 90
Otosclerosis 135
Ovary 31, 37, 44, 51, 97, 123, 128
Oxygen 104, 162
Oxytocin 40, 140

Pachytene 55
Pancreas 121, 126, 145
Pancreozymin 109
Parabiosis 42
Paracrines 46, 47, 121, 145
Parathormone, parathyroid 90, 117, 121, 125, 126
Parkinson's disease 131
Paroophoron 32
Parturition 152
Pedigrees 10
Pelt 94
Peyer's patches 116
Phenotype 9
Pheromones 49, 51, 53, 85, 94, 134, 147
Phosphate 126
Photoperiod 22, 40, 102-113, 133, 139
Phytohaemagglutinin 22, 116

Pigmentation 95, 136, 137
Pineal body 40, 47, 49, 100, 108, 110, 121, 124, 139
Pinealectomy 50, 108, 124
Pituitary 36, 39, 40, 44, 81, 100, 101, 121-124, 131, 161
   adenohypophysis 122
      adenomas 123
      castration cells 123
      hormones: ACTH 40, 108, 109, 115, 121, 123, 127, 128, 133, 140
        FSH 40, 43, 51, 55, 56, 109, 122-124, 133, 159
        growth hormone 40, 81, 94, 109, 121, 123, 133, 140, 159
        ICSH (LH) 52
        LH 43, 51, 52, 55, 56, 107, 109, 121, 123, 124, 133, 157
        MSH 123, 136, 137
        prolactin 40, 45, 54, 107, 109, 121, 133, 140
        TSH 40, 106, 107, 109, 121-123, 133, 135
   grafts 45, 123, 132, 133
   hypophysectomy 41, 44, 45, 124, 138
   neurohypophysis 40
      hormones: ADH 40, 104, 124, 129
        oxytocin 124, 140
   portal system 39, 122
Plasma cells 116
Plasmid 3
Polar body 25, 27
Poliomyelitis 69
Polycythaemia 105, 146
Polygenic effects 69
Polyoestrous 53, 54
Polypeptides 46
Polysaccharides 117
Postsynthetic gap 20
Prednisone 140
Pregnancy (gestation) 39, 52, 54, 76, 82, 95, 131, 133, 137, 156, 159
Prenatal losses 60
Presynthetic gap 20
Progesterone 7, 42, 43, 122, 126, 131, 155
Prokaryocyte 3
Prostaglandins 47, 111, 121
Prostate 31, 33, 128
   cancer 70, 71
Protein 36, 105, 122, 128, 146
Protist 3
Puberty 36, 39, 40, 42, 48-55, 69, 78-82, 100, 131, 137, 138, 141, 155, 158, 159
Purines 69
Pyloric stenosis 66, 67

Index

Quinacrine 61

Receptors 36, 121, 122, 128, 132
Regaud's bodies 29
Releasing factors 40, 45, 80, 132-135, 156
Renin 119, 127, 147
Reserpine 134
Respiratory tract 141, 142
Respiratory diseases 142
Rheumatic chorea 69
Rheumatoid factor 118
Rosette formation 116

Scoliosis 90
Scrotum 53
Schizophrenia 131
Sebaceous glands 94, 96, 97, 121, 122, 136, 139, 140, 142, 143
Secretin 109
Seminal vesicle 31, 33
Seminiferous tubules 52
Sense organs 134, 135
Sertoli cells 37, 51, 52, 55
Shetland ponies 76
Shire horses 76
Sex determination 30-38
    maturation and decline 48-57
    primary ratio 58-62
    reversal 5, 6
    secondary ratio 58-62
        blood group AB 62
        first cousins 62
        mice 62
        Samaritans 62
    tertiary ratio 62-71
Shape of body 89-96
Skeleton 90, 91
Skin and appendages 94-96, 133, 136-140
Spermatogenesis 6, 27, 28, 48, 50-52, 54-56, 97, 98
Spina bifida 66, 67
Spindle apparatus 12, 19
Spleen 44, 98-100, 114-116, 119
Spondylitis 91
Steroids 121, 140
Subcutis 94, 136
Sulfanilamide 112
Sweat glands 94, 136, 139
Synapsis 40

Teeth 142, 143
Teleangiectasy 95
Teleosts 5
Temperature:
    biorhythms 103
    core of body 103, 104
    shell 103, 104

Teratogens 66
Testicular feminization 35, 36
Thermoregulation 102, 103, 141, 142
    central control 104, 105
    circulation of blood 104
    insulation 104
    photoperiod 105
    shivering 104
    sweating 104
Thymosin 117
Thymidine 8, 13
Thymus 114-119, 121, 126, 128, 133, 161
Thyroid, thyroxine 82, 97, 112, 121-125, 138, 159
Tonsils 116
Tortoises 152
Trachea 141
Translocations 28, 154, 158
Trisomy 16, 28, 56, 66, 67, 157
Trophoblast 54
Trout 6
Tunica albuginea 31-33
Tyrosin 110

Udder 94
Uric acid 69
Urogenital tract 146, 147
Uterus 43
    cancer 70, 71
Utriculus masculinus 31

Vagina 43, 50
Vascular system 119
Vellus 137, 138
Verbal formulations 130
Vessels of skin 95
    transposition 67
Virilism 127
Visuospatial skills 130
Vitamin:
    A 49
    B12 162
    D 125, 147
    E 49
Vocal cords 141
Vole 16

Weight 75-85
    neonatal 76
    postnatal 77
    pubertal 78-85
    and life span 82-85
    of organs 90, 96-101
Whooping cough 69
Wolffian body and duct 30-34
Wood lemming 16, 154

THE WYKEHAM SCIENCE SERIES

1	*Elementary Science of Metals*	J. W. Martin and R. A. Hull
2	†*Neutron Physics*	G. E. Bacon and G. R. Noakes
3	†*Essentials of Meteorology*	D. H. McIntosh, A. S. Thom and V. T. Saunders
4	*Nuclear Fusion*	H. R. Hulme and A. McB. Collieu
5	*Water Waves*	N. F. Barber and G. Ghey
6	*Gravity and the Earth*	A. H. Cook and V. T. Saunders
7	*Relativity and High Energy Physics*	W. G. V. Rosser and R. K. McCulloch
8	*The Method of Science*	R. Harré and D. G. F. Eastwood
9	†*Introduction to Polymer Science*	L. R. G. Treloar and W. F. Archenhold
10	†*The Stars: their structure and evolution*	R. J. Tayler and A. S. Everest
11	*Superconductivity*	A. W. B. Taylor and G. R. Noakes
12	*Neutrinos*	G. M. Lewis and G. A. Wheatley
13	*Crystals and X-rays*	H. S. Lipson and R. M. Lee
14	†*Biological Effects of Radiation*	J. E. Coggle and G. R. Noakes
15	*Units and Standards for Electromagnetism*	P. Vigoureux and R. A. R. Tricker
16	*The Inert Gases: Model Systems for Science*	B. L. Smith and J. P. Webb
17	*Thin Films*	K. D. Leaver, B. N. Chapman and H. T. Richards
18	*Elementary Experiments with Lasers*	G. Wright and G. Foxcroft
19	†*Production, Pollution, Protection*	W. B. Yapp and M. I. Smith
20	*Solid State Electronic Devices*	D. V. Morgan, M. J. Howes and J. Sutcliffe
21	*Strong Materials*	J. W. Martin and R. A. Hull
22	†*Elementary Quantum Mechanics*	Sir Nevill Mott and M. Berry
23	*The Origin of the Chemical Elements*	R. J. Tayler and A. S. Everest
24	*The Physical Properties of Glass*	D. G. Holloway and D. A. Tawney
25	*Amphibians*	J. F. D. Frazer and O. H. Frazer
26	*The Senses of Animals*	E. T. Burtt and A. Pringle
27	†*Temperature Regulation*	S. A. Richards and P. S. Fielden
28	†*Chemical Engineering in Practice*	G. Nonhebel and M. Berry
29	†*An Introduction to Electrochemical Science*	J. O'M. Bockris, N. Bonciocat, F. Gutmann and M. Berry
30	*Vertebrate Hard Tissues*	L. B. Halstead and R. Hill
31	†*The Astronomical Telescope*	B. V. Barlow and A. S. Everest
32	*Computers in Biology*	J. A. Nelder and R. D. Kime
33	*Electron Microscopy and Analysis*	P. J. Goodhew and L. E. Cartwright
34	*Introduction to Modern Microscopy*	H. N. Southworth and R. A. Hull
35	*Real Solids and Radiation*	A. E. Hughes, D. Pooley and B. Woolnough
36	*The Aerospace Environment*	T. Beer and M. D. Kucherawy
37	*The Liquid Phase*	D. H. Trevena and R. J. Cooke
38	†*From Single Cells to Plants*	E. Thomas, M. R. Davey and J. I. Williams
39	*The Control of Technology*	D. Elliott and R. Elliott
40	*Cosmic Rays*	J. G. Wilson and G. E. Perry
41	*Global Geology*	M. A. Khan and B. Matthews
42	†*Running, Walking and Jumping: The science of locomotion*	A. I. Dagg and A. James
43	†*Geology of the Moon*	J. E. Guest, R. Greeley and E. Hay
44	†*The Mass Spectrometer*	J. R. Majer and M. P. Berry
45	†*The Structure of Planets*	G. H. A. Cole and W. G. Watton
46	†*Images*	C. A. Taylor and G. E. Foxcroft
47	†*The Covalent Bond*	H. S. Pickering
48	†*Science with Pocket Calculators*	D. Green and J. Lewis
49	†*Galaxies: Structure and Evolution*	R. J. Tayler and A. S. Everest
50	†*Radiochemistry—Theory and Experiment*	T. A. H. Peacocke
51	†*Science of Navigation*	E. W. Anderson
52	†*Radioactivity in its historical and social context*	E. N. Jenkins and I. Lewis
53	†*Man-made Disasters*	B. A. Turner

†(*Paper and Cloth Editions available.*)

THE WYKEHAM ENGINEERING AND TECHNOLOGY SERIES

1	*Frequency Conversion*	J. THOMSON, W. E. TURK and M. J. BEESLEY
2	*Electrical Measuring Instruments*	E. HANDSCOMBE
3	*Industrial Radiology Techniques*	R. HALMSHAW
4	*Understanding and Measuring Vibrations*	R. H. WALLACE
5	*Introduction to Tribology*	J. HALLING and W. E. W. SMITH

All orders and requests for inspection copies should be sent to the appropriate agents. A list of agents and their territories is given on the verso of the title page of this book.

2961-4
5-14